WIDEBAND LOW NOISE AMPLIFIERS EXPLOITING THERMAL NOISE CANCELLATION

THE KLUWER INTERNATIONAL SERIES IN ENGINEERING AND COMPUTER SCIENCE

ANALOG CIRCUITS AND SIGNAL PROCESSING

Consulting Editor*: Mohammed Ismail. *Ohio State University

Related Titles:

SYSTEMATIC MODELING AND ANALYSIS OF TELECOM FRONTENDS AND THEIR BUILDING BLOCKS
Vanassche, Gielen, Sansen
ISBN: 1-4020-3173-4

LOW-POWER DEEP SUB-MICRON CMOS LOGIC SUB-THRESHOLD CURRENT REDUCTION
van der Meer, van Staveren, van Roermund
ISBN: 1-4020-2848-2

OPERATIONAL AMPLIFIER SPEED AND ACCURACY IMPROVEMENT
Ivanov and Filanovsky
ISBN: 1-4020-7772-6

STATIC AND DYNAMIC PERFORMANCE LIMITATIONS FOR HIGH SPEED D/A CONVERTERS
van den Bosch, Steyaert and Sansen
ISBN: 1-4020-7761-0

DESIGN AND ANALYSIS OF HIGH EFFICIENCY LINE DRIVERS FOR Xdsl
Piessens and Steyaert
ISBN: 1-4020-7727-0

LOW POWER ANALOG CMOS FOR CARDIAC PACEMAKERS
Silveira and Flandre
ISBN: 1-4020-7719-X

MIXED-SIGNAL LAYOUT GENERATION CONCEPTS
Lin, van Roermund, Leenaerts
ISBN: 1-4020-7598-7

HIGH-FREQUENCY OSCILLATOR DESIGN FOR INTEGRATED TRANSCEIVERS
Van der Tang, Kasperkovitz and van Roermund
ISBN: 1-4020-7564-2

CMOS INTEGRATION OF ANALOG CIRCUITS FOR HIGH DATA RATE TRANSMITTERS
DeRanter and Steyaert
ISBN: 1-4020-7545-6

SYSTEMATIC DESIGN OF ANALOG IP BLOCKS
Vandenbussche and Gielen
ISBN: 1-4020-7471-9

SYSTEMATIC DESIGN OF ANALOG IP BLOCKS
Cheung & Luong
ISBN: 1-4020-7466-2

LOW-VOLTAGE CMOS LOG COMPANDING ANALOG DESIGN
Serra-Graells, Rueda & Huertas
ISBN: 1-4020-7445-X

CIRCUIT DESIGN FOR WIRELESS COMMUNICATIONS
Pun, Franca & Leme
ISBN: 1-4020-7415-8

DESIGN OF LOW-PHASE CMOS FRACTIONAL-N SYNTHESIZERS
DeMuer & Steyaert
ISBN: 1-4020-7387-9

MODULAR LOW-POWER, HIGH SPEED CMOS ANALOG-TO-DIGITAL CONVERTER FOR EMBEDDED SYSTEMS
Lin, Kemna & Hosticka
ISBN: 1-4020-7380-1

DESIGN CRITERIA FOR LOW DISTORTION IN FEEDBACK OPAMP CIRCUITE
Hernes & Saether
ISBN: 1-4020-7356-9

CIRCUIT TECHNIQUES FOR LOW-VOLTAGE AND HIGH-SPEED A/D CONVERTERS
Walteri
ISBN: 1-4020-7244-9

DESIGN OF HIGH-PERFORMANCE CMOS VOLTAGE CONTROLLED OSCILLATORS
Dai and Harjani
ISBN: 1-4020-7238-4

CMOS CIRCUIT DESIGN FOR RF SENSORS
Gudnason and Bruun
ISBN: 1-4020-7127-2

WIDEBAND LOW NOISE AMPLIFIERS EXPLOITING THERMAL NOISE CANCELLATION

by

Federico Bruccoleri

Catena Microelectronics B.V.,
Delft, The Netherlands

Eric A.M. Klumperink

University of Twente,
Enschede, The Netherlands

and

Bram Nauta

University of Twente,
Enschede, The Netherlands

Springer

A C.I.P. Catalogue record for this book is available from the Library of Congress.

ISBN 1-4020-3187-4 (HB)
ISBN 1-4020-3188-2 (e-book)

Published by Springer,
P.O. Box 17, 3300 AA Dordrecht, The Netherlands.

Sold and distributed in North, Central and South America
by Springer,
101 Philip Drive, Norwell, MA 02061, U.S.A.

In all other countries, sold and distributed
by Springer,
P.O. Box 322, 3300 AH Dordrecht, The Netherlands.

Printed on acid-free paper

Printed in the Netherlands.

Preface

It was around the mid-nineties at the University of Twente, when Eric Klumperink had a wild idea: since popular analog basic circuits, like the differential pair, current mirror etc are all very simple, he wondered if we already know all possible simple circuits. Starting with two MOSFETs and resistors, he tried to figure out how many circuits he could generate. As a transistor usually acts as a voltage controlled current source, while resistors can also be modeled in that way, Eric decided to use voltage controlled current sources as building blocks and find two-port circuits with a non-zero transfer function. Using brains, graph theory, and a MAPLE computer program, he found them all: 145 circuit-graphs with two voltage controlled current sources. As each graph has several different transistor implementations, hundreds of transistor-resistor circuits are possible. He classified and analyzed circuits in his PhD thesis, finding some with "interesting" thermal noise cancellation, but concluded that it was not trivial to find the "really useful" ones. In december 1997, Federico Bruccoleri accepted a PhD position, and his task was to find "really useful" circuits. He limited himself to wideband Low Noise Amplifers (LNAs) and wrote down the boundary conditions for these LNAs (like wideband gain, well-defined input impedances, etc). He then searched for graphs satisfying the conditions, limiting himself to two-transistor circuits. He found four wideband LNAs: two were known circuits and two were indeed new. Thanks to careful analysis, Federico discovered that one of the new LNAs outperformed the others, because part of the thermal noise of the input device was cancelled! This was verified by chip measurements, and received a best Poster Award at ESSCIRC 2000. This stimulated Federico to generalize the noise cancellation concept, resulting in a class of circuits where the noise of the input device is completely cancelled. That was the moment when we really got excited! A next IC implementation was fabricated, and measurements showed indeed that the thermal noise of the input transistor was completely cancelled. This circuit was presented at the ISSCC 2002 conference, and won the "IEEE 2002 ISSCC Jan van Vessem Outstanding European Paper Award". In my opinion this noise canceling technique creates a new degree of freedom in wide band amplifier design which may turn out to be very useful in future products.

Together with Eric Klumperink, it was a great pleasure to supervise, the PhD work of Federico. This book describes the PhD work and some of Eric's work. I hope you enjoy reading it.

Prof. Bram Nauta,
University of Twente

Preface

Contents

Chapter 1
Introduction

1.1 Introduction

Over the last decade, the penetration of wired and wireless digital communication devices into the mass market has been very pervasive. In order to meet stringent market requirements, low-cost flexible digital-communication systems capable of high data-rates and increased functionality are desired [1]. From the point of view of the hardware, two important consequences can be identified. First, system-on-a-chip (SoC) integration intended as integration of electronic functions that are currently implemented with different IC process technologies is mandatory to reduce further manufacturing costs [1]. In this respect, driven by a reduction in minimum feature size of about 70% each 2 to 3 years, MOS technologies are the most promising for SoC integration. Second, in contrast to circuit techniques exploiting the narrow-band nature of L-C resonant tanks, wide-band circuits are inherently suited to accommodate high data-rates and lend themselves to the realisation of flexible multi-functional communication systems.

In communication systems, electronic transmitter and receiver circuits transfer information to and from a communication medium (e.g.: air). The receiver side presents challenges, which are not present or are greatly relaxed for the transmitter. This is mainly due to the hostile nature of the communication channel, which results in a minimum detectable signal at the receiver input that can be as weak as a few μVolts. The receiver must be able to handle such a signal in order to guarantee a reliable quality of the information transfer. This ability of the receiver to detect a weak input signal (i.e. referred as its sensitivity) is fundamentally limited by the electrical noise present at its input. Specifically, for a given modulation scheme and after decoding (e.g.: de-spreading for systems using direct-sequence spread-spectrum), a certain minimum signal-to-noise ratio SNR is required to achieve the desired bit error rate. Since electrical noise is a fundamental obstacle to the reception of weak signals, low-noise techniques are crucial for receiver design.

This book deals with wide-band high-performance low-noise techniques that:

- Exploit the intrinsically wide-band transconductance of MOSFETs and
- are suitable to be implemented in low-cost highly integrated receiver architectures.

In contrast to other techniques such as distributed amplification [2,3,4] and wide-band matching [5], this work focuses on circuit solutions that do *not* rely on the behaviour of coils or transmission lines in order to achieve a low-noise performance over a wide range of frequencies. Especially in the low GHz range, where most of the radio applications reside, integrated coils and transmission lines require a large chip-area. Moreover, their quality is lower for a low-cost standard CMOS process compared to other dedicated RF technologies like BiCMOS, Si/SiGe Bipolar and GaAs. Instead, this work focuses on circuit techniques that can achieve low-noise behaviour exploiting the wide-band nature of transistors and resistors, which are readily available in any CMOS technology.
Although the circuit techniques presented in this book are described for CMOS, they also can be applied for other technologies like BiCMOS, Bipolar and GaAs.

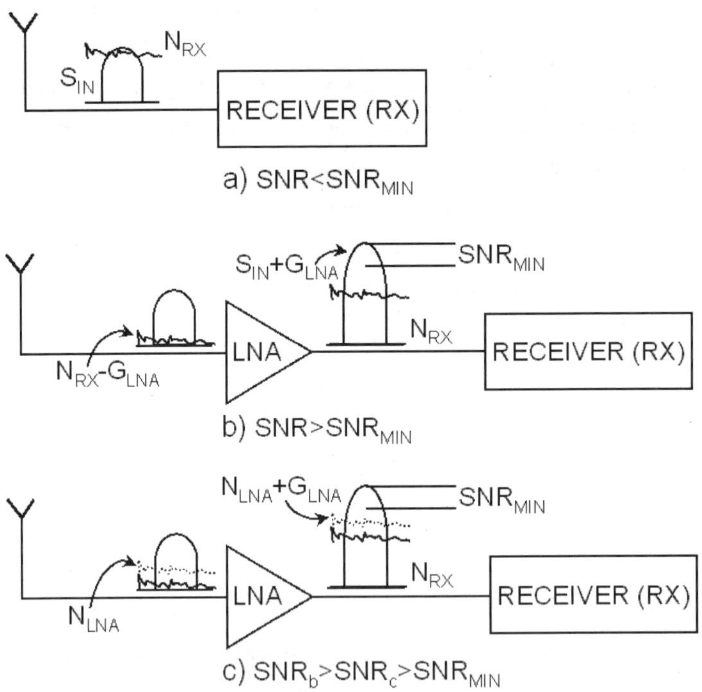

Figure 1.1: Placing a Low-Noise Amplifier in front of a receiver to improve its sensitivity: a) w/o LNA, b) w noiseless LNA and c) w LNA noise included. All the quantities are in dB's.

1.2 Motivation

In the following, some argumentation concerning the motivations of this work is provided.

- Need of a Low-Noise Amplifier

Figure 1.1 illustrates, in qualitative terms, the importance of placing a low-noise amplifier (LNA) in front of a receiving system characterised by a poor input sensitivity. For sake of simplicity, in the following discussion it will be assumed that a certain positive[I] minimum SNR at the receiver input, SNR_{MIN}, is required in order to obtain the desired bit error rate. Furthermore, the signal S, noise N and the gain G are all expressed in unit of dB's.

In figure 1.1a, the minimum detectable signal, S_{IN}, at the input of the receiver is below its input noise-floor, N_{RX}. In this condition, signal reception is hampered because the input SNR is lower than the required SNR_{MIN}. On the other hand, figures 1.1b-c show how signal reception is restored by placing in front to the receiver a low-noise amplifier (LNA) with proper noise and gain characteristics. In figure 1.1b, the signal at the receiver input, $S_{IN}+G_{LNA}$, is brought above the noise-floor N_{RX} by choosing the gain of the LNA G_{LNA} large enough (i.e. such $SNR=S_{IN}+G_{LNA}-N_{RX}>SNR_{MIN}$ holds). In other words, the receiver noise-floor referred to the input of the LNA, $N_{RX}-G_{LNA}$, is now properly small compared to S_{IN} or $N_{RX}-G_{LNA}<S_{IN}-SNR_{MIN}$. In theory, G_{LNA} can be chosen such that no significant RX noise can be referred at the input of the LNA[II]. Clearly, the provision of a sufficiently large forward gain is an essential LNA requirement. For a large G_{LNA}, the noise added by the LNA itself almost entirely determines the noise-floor of the *overall* chain (i.e. LNA plus receiver) and so its sensitivity. This leads to the second fundamental requirement for the LNA: its (equivalent) input noise N_{LNA} must be small to achieve the desired sensitivity (i.e. $N_{LNA}<S_{IN}-SNR_{MIN}$ holds). In this respect, the lower the LNA input noise N_{LNA}, the higher is the ability to detect weak input signals. This has important practical consequences both at system and circuit level. For a given transmitter output power, a larger path-loss is tolerated (i.e., the receiver can be located at a longer distance from the transmitter). In turn, this means that the number of transmitters needed to cover a certain area is lower. Alternatively, a transmitter using less output power can accommodate the same distance from the receiver. From circuit-design point of view a lower N_{LNA} relaxes the demand over the LNA gain and/or the noise generated in the following stages. Lower LNA gain means a smaller signal and so relaxes its linearity requirements for the

[I] In communication systems exploiting direct-sequence spread-spectrum techniques, the SNR at the antenna is allowed to be *negative* but above a system-dependent minimum, $SNR_{MIN}<0$. The latter is required to guarantee a *positive* SNR_{MIN} at the demodulator output *after* de-spreading.

[II] The LNA gain is limited to some maximum value, for instance due to the non-linearity of the following stage.

following stage. From the above discussion, it follows that for radio applications requiring high-sensitivity, a LNA is always the *first* active block in the receiver chain.

- Need of a Wide Operation Bandwidth

Wide-band Low-Noise Amplifiers (LNAs) essentially built of transistors and resistors are commonly found in receiving systems where the ratio between the application bandwidth BW and its middle frequency[III] $BW/(f_l+BW/2)$ can be as large as two, as shown in figure 1.2. Examples are analogue cable (BW: 50-850MHz), satellite (BW: 950-2150MHz) and terrestrial (BW: 450-850MHz) digital video broadcasting. At these frequencies, parallel and series L-C tanks used as load and for source degeneration do not perform adequately over the bandwidth or integrated coils tend to be too large and so bulky. However, a wide-band inductor-less LNA can alternatively replace multiple parallel LC-tuned LNAs traditionally employed in multi-band [6-11] and multi-mode [12] narrow-band receivers to accommodate more frequency bands *once at the time* or *simultaneously*, respectively.

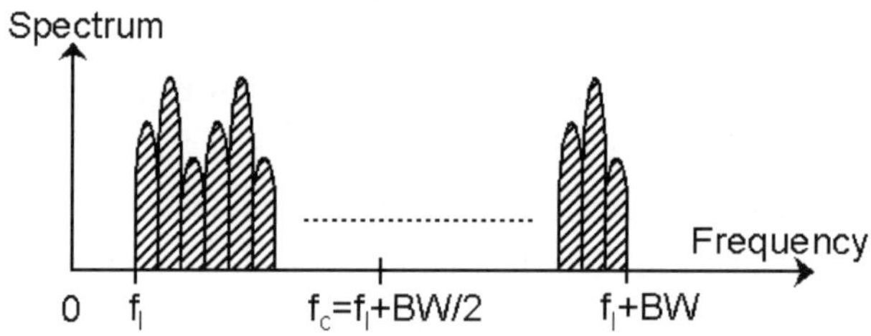

Figure 1.2: A wide-band signal spectrum spanning a bandwidth from f_l to f_l+BW with centre frequency, $f_l+BW/2$, such that the ratio $BW/(f_l+BW/2)$ can be as large as 2 (i.e. for $BW/2>>f_l$).

In this case, a wide-band LNA can save significant chip-area because integrated inductors typically are the most area consuming among on-chip components. Furthermore, a wide-band LNA provides an increased flexibility of the front-end, a step-forward in the direction of a programmable software-defined radio. Finally, for multi-mode operation, the typically larger power dissipation of a wide-band amplifier is a minor issue because the sum of the power dissipations of the tuned LNAs is made available. This last point does not hold for a concurrent multi-band LNA design [13] because in this case a single active component is combined to a matching network with multiple-resonance covering the different frequency bands. Nevertheless, the quality factor of on-chip inductors, especially in MOS processes, limits the capability to accommodate contiguous frequency bands.

[III] This ratio is sometime referred as fractional bandwidth.

1.3 Outline

This work aims to investigate alternative wide-band low-noise circuit techniques suitable for low-cost high-performance integrated MOS receivers. This goal is pursued seeking a radically different design approach. Exploiting the fact that many elementary wide-band amplifiers exploit the wide-band transconductance of the MOS transistor to determine their performance, in chapter 2 *all* amplifiers that can be modelled as circuits with 2 Voltage Controlled Current Source (VCCS) are generated in a systematic fashion. To do so, a methodology is exploited, which renders *all* 2VCCS wide-band amplifiers that can be found in a database containing *all* the potentially useful 2VCCS circuits, which was made available from previous work of Klumperink [14,15, appendix D]. This is done selecting into the 2VCCS database *all* two-port circuits having certain non-zero transmission parameters {A, B, C, D} according to a set of properly defined amplifier *functional* requirements and given the source/load impedance. Limiting ourselves to the important case of elementary 2VCCS circuits exploiting 2 MOSFETs leads to 2 *well-know* and 2 *unknown* wide-band amplifier circuits. In chapter 3, the small-signals and noise properties of the new wide-band amplifiers are analysed. The purpose is to compare their properties with that of known circuits. It will be shown that for one of the new amplifiers a useful noise cancellation mechanism occurs, which leads to superior noise performance for a given voltage gain, power dissipation and upon source impedance matching. This behaviour was verified through the design and the experimental measurements of a 0.35μm 50 to 900MHz wide-band variable-gain LNA. Focusing on low-noise, chapter 4 begins with a review of properties and limitations of commonly used MOS wide-band techniques. From the understanding of their fundamental properties and limitations, a novel wide-band low-noise technique is then proposed, which is an improved version of the noise cancellation mechanism described in chapter 3. By exploiting this noise cancellation technique, wide-band LNAs can be designed to provide an arbitrarily low NF upon source impedance matching without suffering from instability issues typical of commonly used wide-band amplifiers exploiting global negative feedback. The noise cancellation technique is then generalised to other two-port circuit implementations and its basic properties (i.e. simultaneous noise-power match, distortion cancellation and robustness to device parameter variations) and high-frequency limitations are analysed in. In chapter 5, the noise cancelling theory is validated through the design and experimental verification of a decade-bandwidth sub-2.4dB NF LNA in 0.25μm CMOS. Chapter 6 presents conclusions and recommendations for further research.

1.4 References

[1] Dennis D. Buss, "Technology in the Internet Age", IEEE 2002 International Solid-State Circuits Conference, digest of technical paper, SF, California, USA.

[2] T. T. Y. Wong, "Fundamentals of Distributed Amplification", Norwood, Artech, 1993.

[3] B. M. Ballweber, R. Gupta, and D. J. Allstot, "A Fully Integrated 0.5-5.5-GHz CMOS Distributed Amplifier", IEEE J. of Solid-State Circuits, vol. 35, pp. 231-239, Feb. 2000.

[4] H. T. Ahn and D. J. Allstot, "A 0.5-8.5-GHz Fully Differential CMOS Distributed Amplifier", IEEE Journal of Solid-State Circuits, vol. 37, pp. 985-993, August. 2002.

[5] D. M. Pozar, "Microwaves Engineering", Second Edition, Wiley, 1998.

[6] S. Wu and B. Razavi,"A 900MHz/1.8GHz CMOS receiver for dual-band applications", IEEE Journal of Solid-State Circuits, vol.33, pp.2178-2185, Dec. 1998.

[7] J. Tham et al, "A 2.7V 900-MHz dual-band transceiver IC for digital wireless communication", IEEE Journal of Solid-State Circuits, vol.34, pp.286-291, Mar. 1999.

[8] J. Imbornone et al, "fully Differential dual-band image reject receiver in SiGe BiCMOS", in IEEE RFIC Symposium Digest, 2000, pp. 147-150.

[9] W. Titus et al, "A Si BJT dual-band receiver IC for DAB", in IEEE microwave millimiter-Wave Monolithic Circuits Symposium Digest, vol.1, pp.345-348, 1998.

[10] S. Dow et al, "A Dual-Band Direct-Conversion/VLIF Transceiver for 850GSM/ GSM/DCS/PCS", IEEE International Solid-State Circuits Conference, 2002.

[11] A. Molnar et al, "A Single-Chip Quad-band Direct-Conversion GSM/GPRS RF Transceiver with Integrated VCOs and Fractional-N Synthesizer", IEEE International Solid-State Circuits Conference, digest of technical paper, 2002.

[12] V. Aparin et al, "A highly-Integrated tri-Band/Quad-Mode SiGe BiCMOS RF-to-Baseband Receiver for Wireless CDMA/WBCDMA/AMPS Applications with GPS Capability", IEEE International Solid-State Circuits Conference, digest of technical paper,

[13] H. Hashemi and A. Hajimiri, "Concurrent Multiband Low-Noise Amplifiers - Theory, design and Applications", Invited Paper on IEEE Transactions On Microwaves Theory and Techniques, vol.50, NO. 1, Jan. 2002.

[14] E.A.M. Klumperink, "Transconductance based CMOS Circuits: Generation, Classification and Analysis", PhD. Thesis, University of Twente, Enschede, The Netherlands, 1997.

[15] E.A.M. Klumperink, et al, "Finding all the elementary circuits exploiting transconductance", IEEE Transaction on Circuits and Systems, CAS-II, vol. 48, No. 11, pp. 1039-1053, Nov. 2001.

Chapter 2
Systematic Generation of *All* Elementary Wide-Band Amplifiers

2.1 Introduction

Designers, generally conceive new amplifier circuits by exploiting their creativity, insight and experience. As this is a largely unstructured process, it is unlikely that *all* the useful amplifier alternatives are found. In contrast, this chapter describes a radically different approach that generates a*ll* potentially useful alternatives to well-known elementary wide-band amplifiers like the common-gate and the common-source shunt-feedback stage. This is done by defining a methodology that generates systematically *all* the two-port amplifiers that can be modelled as circuits with 2 Voltage Controlled Current Sources (VCCS). Important reasons to exploit a VCCS as circuit generating element are:

- The small-signal operation of a MOSFET -in saturation- is essentially that of a linear VCCS element "I=g·V" with "g" equal to the device transconductance "g_m". This simplified model is valid for frequencies where the non quasi-static effects of the device are negligible [1]: up to tens of GHz for a deep sub-micron CMOS process.
- Commonly used elementary amplifiers [2] such as the common gate, common drain and the common source shunt-feedback stages exploit the "g_m" of a MOST to define their small-signal transfer properties like gain and port impedances. Their functional behaviour is adequately represented when regarding them as circuits with 1 VCCS or with 2 VCCSs (i.e. 1VCCS or 2VCCS circuits).
- Several different transistor circuits are automatically covered because a 4-terminal VCCS plus extra interconnections can model any combinations of MOSFETs and resistors acting as a transconductor circuit as well as a simple resistor as illustrated in figure 2.1.

Furthermore, at least 2 VCCSs are required in order to provide voltage (or current) gain larger than one. Next, as will become clear later in this chapter, the use of only 2VCCS means that the achievements of previous work can be directly exploited.

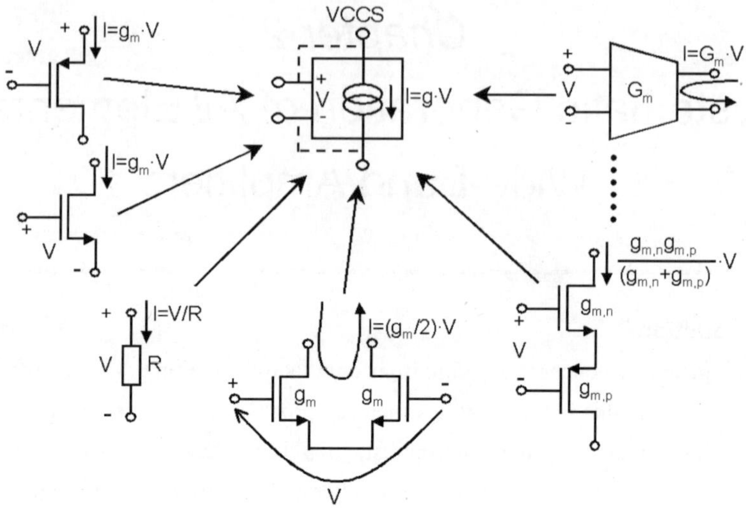

Figure 2.1: A VCCS can model a single MOSFET, a single resistor or any transconductor circuit.

2.2 The Systematic Generation Methodology

As previously mentioned, the aim of this chapter is to find all the potentially useful alternatives to well-known elementary wideband amplifiers by generating systematically all the two-port amplifiers that can be seen as circuits with 2 VCCSs. In previous work of Klumperink [3,4,5], all the graphs of potentially useful two-port circuits built by the interconnection of 2 VCCSs were systematically generated, classified in terms of their properties and stored in a database: the *2VCCS graph database*. This research starts from the fact that all the graphs of 2VCCS wideband amplifiers must be a sub-set of the 2VCCS database. In order to find them all, a systematic selection procedure will be described. In the next sub paragraph, the generation of the 2VCCS database and its properties are briefly reviewed. More details can be found in [3, 4, 5 and appendix D]. Following, the systematic selection of all the graphs of 2VCCS wideband amplifiers is described in detail.

2.2.1 2VCCS graphs database: generation and properties [3, 4 and 5]

The flow chart in figure 2.2a describes the steps leading to the generation of the 2VCCS graph database. In facing the problem of how to generate all the topologies of two-port circuits with 2 VCCSs a complexity issue pops up immediately. Since a VCCS has 4 terminals, a lot of circuits are possible interconnecting two of them (i.e. $VCCS_a$ and $VCCS_b$). This is evident even if one considers the simple case of a two-port connected between a voltage source V_S with impedance Z_S and load impedance Z_L (figure 2.2b).

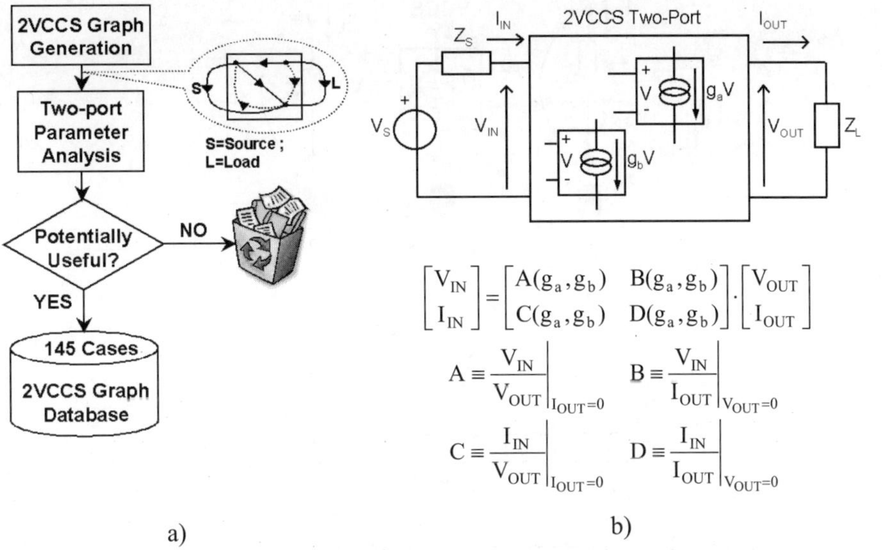

Figure 2.2: a) Flow chart describing the steps leading to the generation of the 2VCCS database b) A 2VCCS two-port circuit described by {A, B, C, D} parameters. The latter are a function of the transconductances g_a and g_b of the 2 VCCSs (i.e. $VCCS_a$ and $VCCS_b$).

This issue was successfully solved [3, 4, 5] by considering that the connectivity among VCCSs can be efficiently studied representing the VCCS by a graph with a "V" and "I" branch as shown in figure 2.3a. The graphs of the source "S" and the load "L" are also shown. Graph theory could then be applied in order to generate systematically all the possible graphs of 2VCCS circuits avoiding to find the same one twice. This rendered 16000 2VCCS graphs! The small-signal properties of two-port circuits of these graphs were then analysed looking at their two-port parameters. To this purpose, {A, B, C, D} transmission parameters (see their definition in figure 2.2b) were preferred to other types of parameters because they are related to the transfers of a two-port: A=1/VoltageGain, B=1/Transconductance, C=1/Transimpedance and D=1/CurrentGain. If potentially useful 2VCCS graphs have at least *one non-zero* {A, B, C, D} parameter, we are left with only 145 graphs, which are stored in a database. Figure 2.3b gives an overview of the combinations of {A, B, C, D} parameters in the 2VCCS database, revealing a great deal of cases. Possible expressions for the {A, B, C, D} parameters as a function of the transconductances g_a and g_b of the 2 VCCSs and the number of graphs are indicated. In the next paragraph, the information in figure 2.3b will be used to select all graphs of wideband amplifiers in the 2VCCS database.

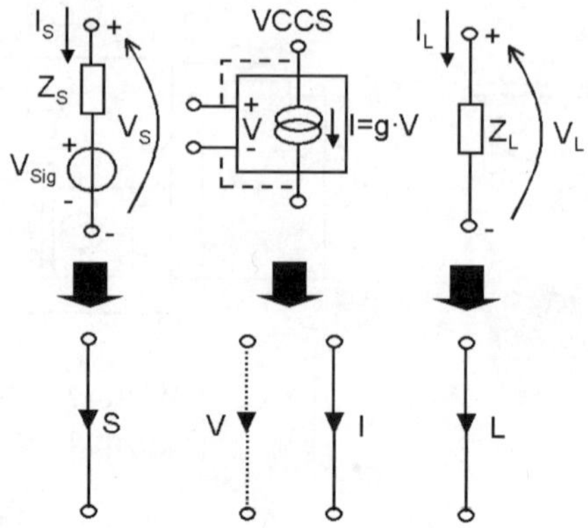

a)

CASE	A	B	C	D	Nr.
{A}	1	0	0	0	3
{B}	0	$1/g_1$	0	0	37
{D}	0	0	0	1	2
{AB}	1 or g_1/g_2	$1/g_3$	0	0	24
{AD}	1	0	0	1	6
{BC}	0	$1/g_1$	g_2	0	2
{BD}	0	$1/g_1$	0	1 or g_2/g_3	24
{ABC}	1 or g_1/g_2	$1/g_3$	g_4	0	3
{ABD}	1 or g_1/g_2	$1/g_3$	0	1 or g_4/g_5	24
{ACD}	1	0	g_1	1	9
{BCD}	0	$1/g_1$	g_2	1 or g_2/g_3	4
{ABCD}	1 or g_1/g_2	$1/g_3$	g_4	1 or g_5/g_6	7

b)

Figure 2.3: a) The VCCS, the source V_S and the load impedance Z_L represented via graphs: "V" and "I" branch for the VCCS, "S" branch for the source and a "L" branch for the load b) Combinations of {A, B, C, D} parameters available in the 2VCCS database. Parameter expressions as a function of g_a and g_b and number of graphs (Nr.) are shown.

2.3 Functional Selection of All Elementary Wide-Band Amplifiers

This paragraph describes the basic idea exploited to select all the graphs of wideband amplifiers within the 2VCCS database. In general, the input impedance Z_{IN}, output impedance Z_{OUT}, forward voltage gain A_{VF} and reverse voltage gain A_{VR} of any linear two-port circuit is a function of the two-port parameters, the source impedance Z_S and the load impedance Z_L. This is shown by the following two-port equations using {A, B, C, D} parameters:

$$Z_{IN} = \frac{Z_L A + B}{Z_L C + D} \qquad Z_{OUT} = \frac{Z_S D + B}{Z_S C + A}$$
$$A_{VF} \equiv \frac{V_{OUT}}{V_{IN}} = \frac{Z_L}{Z_L A + B} \qquad A_{VR} \equiv \frac{V_{IN}}{V_{OUT}} = \frac{AD - BC}{D + B/Z_S} \tag{2.1}$$

Equations (2.1) show that ports impedance (i.e. Z_{IN}, Z_{OUT}) and gains (i.e. A_{VF} and A_{VR}) are univocally determined by the {A, B, C, D} parameters, Z_S and Z_L as shown in figure 2.4a.

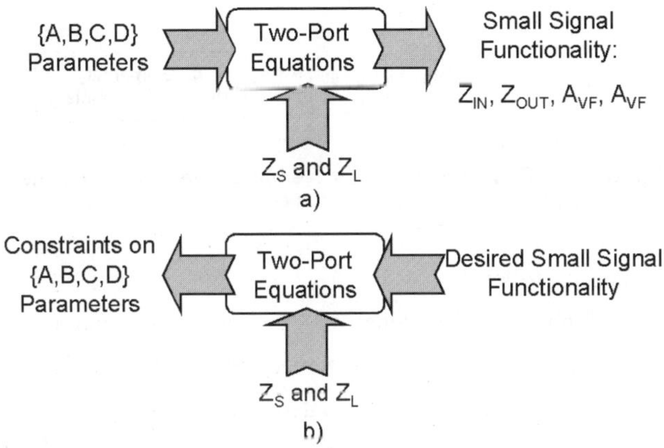

Figure 2.4: a) {A, B, C, D} parameters plus Z_S and Z_L determine the two-port small-signal functionality. b) Graphs of 2VCCS wideband amplifiers are selected using the reverse reasoning: the desired amplifier functionality is traduced into constraints on the {A, B, C, D} parameters upon assigned Z_S and Z_L. These constraints are the criteria to select useful graphs within the 2VCCS database.

The procedure developed to select systematically all the graphs of 2VCCS wideband two-port amplifiers exploit the reverse reasoning indicated in figure 2.4b. In this case, the behaviour of a wideband two-port amplifier is first defined in terms of proper *functional* requirements, which are then translated into constraints for the {A, B, C, D} parameters of two-port circuits. To do so, equations (2.1) are exploited upon properly defined source and load impedance. The derived constraints are finally used as criteria to select graphs of wideband amplifiers in the 2VCCS database. This selection procedure has been implemented into the 4-step procedure described in the flowchart of figure 2.5.

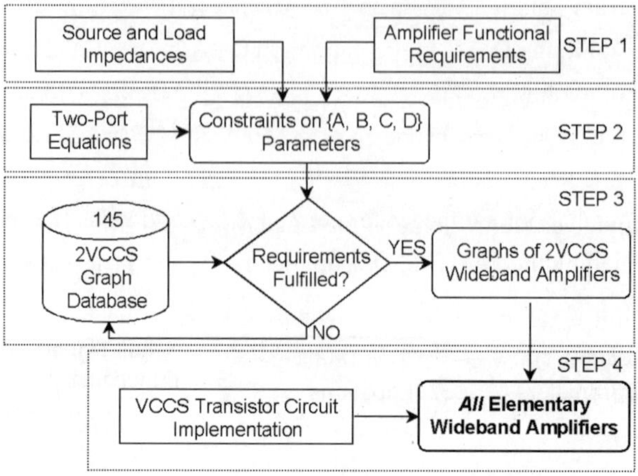

Figure 2.5: Flowchart of the systematic selection of all the elementary two-port amplifiers.

The procedure proceeds as follows:

STEP1: Amplifier high-level functional requirements and source/load impedances suitable for highly integrated CMOS receivers are defined.

STEP2: The information of STEP1 is translated into constraints for the {A, B, C, D} parameters of two-port amplifiers.

STEP3: Graphs of two-port circuits that meet these constraints are selected within the 2VCCS database.

STEP4: MOS transistor circuits of the selected graphs are provided.

In the next sections, the above steps are described.

2.3.1 STEP1: Source/load impedance and functional requirements

Source/load impedance and the amplifier functional requirements crucially determine the output of the generation methodology. For a wideband amplifier to be used in the receiver front-end of a modern communication system, the following assumptions about the source impedance Z_S and load impedance Z_L are made:

- The source impedance Z_S is *real*: Z_S=R_S. This choice is motivated by the fact that the off-chip signal source represents either a coaxial cable terminated on its characteristic impedance or a discrete RF filter, which also provide a real output impedance (at least within a specified range of frequencies). Typical values of R_S are 50Ω and 75Ω.
- The load impedance Z_L is *capacitive*: Z_L=1/(jωC_L). This choice is dictated by the realization of highly integrated receivers exploiting architectures with a minimum

number of external components. In this respect, zero-IF and low-IF receivers are the most indicated solutions [6, 7]. In these architectures, the front-end amplifier is directly loaded by the following on-chip frequency mixer (i.e. the external image rejection filter is not required). In principle, the choice of the input impedance of the mixer is a degree of freedom. In practice, commonly used active mixer circuits (i.e. the so-called Gilbert-type mixer) provide capacitive input impedance [6, 7].

The following requirements on the functionality of a wide-band amplifier are important:

- **Gain.** The amplifier must provide sufficient forward voltage gain: $|A_{VF}|>1$. This is required in order to boost a weak input signal above the generally high input noise-floor of the following frequency mixer. Moreover, the reverse gain A_{VR} must be low enough to isolate the amplifier input from any undesired signal injected at its output.
- **Source impedance matching.** The amplifier input impedance must match the source impedance R_S: $Z_{IN}=R_S$[IV]. Incorrect termination of a coaxial cable leads to signal reflections that can cause destructive interference at the amplifier input. Incorrect termination of the RF filter preceding the amplifier leads to alterations of its transfer characteristics such as in-band ripples (even notches) and poorer out-band attenuation [8]. Signal reflections and in-band ripples degrade the receiver sensitivity while poorer out-band attenuation leads to receiver overloading.
- **Stability.** The amplifier must be stable at all the frequencies and upon all operating conditions. This includes (a) device parameter variations due to process-spread and temperature, (b) inaccurate or lacking modelling for the active devices, substrate underneath, IC package, and source/load impedances and (d) large signal operation. To cope with these issues, unconditional stability is typically required, which provides the safest degree of stability [9].
- **Frequency behaviour.** The frequency response of an amplifier is assumed wide-band if its transfer functions are frequency-independent in $[f_l, f_l+BW]$ and the ratio between the bandwidth BW and its middle frequency, $BW/(f_l+BW/2)$, can be as large as 2.

A step-up 1:n transformer with a resistive output termination equal to $n^2 \cdot R_S$ meets the above requirements (i.e. $Z_{IN}=R_S$, $A_{VF}=n$ and $A_{VR}=1/n$). However, transformers are not considered because they require a large area while their wide-band performance is typically poor, especially in CMOS processes, and anyhow at frequencies below one GHz.

[IV] A certain mismatch is tolerated. Typical values of $|\Gamma_{IN}|=|(Z_{IN}-R_S)/(Z_{IN}+R_S)|$ are from –8dB to –10dB.

2.3.2 STEP2: Constraints on the two-port {A, B, C, D} parameters

In this section, general constraints for the {A, B, C, D} parameters of two-port circuits are derived using the defined functional requirements and equations (2.1). Two types of constraints are distinguished: 1) on the allowed combinations of {A, B, C, D} parameters and 2) on the value of the non-zero {A, B, C, D} parameters.

Allowed combinations of {A, B, C, D} parameters. The two-port equations (2.1) suggest that not all the combinations of {A, B, C, D} parameters can be used to implement the functionality of a wideband amplifier. In table 2.1, expressions for the two-port input impedance Z_{IN} and the forward gain A_{VF} are given for all the combinations of {A, B, C, D} parameters. All two-ports with one non-zero transmission parameter and two-ports {AB} and {CD} are useless for our purposes because they render a Z_{IN} that is either 0 or ∞. For the remaining cases, further selection is done analysing the qualitative behaviour of Z_{IN} and A_{VF} versus frequency due to $Z_L=1/(j\omega C_L)$ as shown in figures 2.6a and 2.6b. For instance, two-ports {AD} and {BC} are useless as their Z_{IN} is imaginary and strongly frequency-dependent through Z_L (i.e. integrative and derivative frequency behaviour, see also table 2.1). For the remaining two-port cases {{AC}, {AB}, {BD}, {ABC}, {ABD}, {ACD}, {BCD}, {ABCD}}, a wide range of frequencies [f_l, f_l+BW] can be found in figure 2.6a, where a real Z_{IN} can be made equal to R_S. However, cases {{BD}, {BCD}} are rejected because their gain A_{VF} has an integrative response (figure 2.6b). Case {ABD} is rejected because it leads to conflicting demands on Z_{IN} and A_{VF} (i.e. from table 2.1, a wideband Z_{IN} is requires $|Z_L\cdot A|<<|B|$ while a wideband A_{VF} requires $|Z_L\cdot A|>>|B|$). Ultimately, wideband two-port amplifiers must have one of the following combinations of non-zero {A, B, C, D} parameters: {{AC}, {ABC}, {ACD}, {ABCD}}. Note as parameters “A” and “C” are always present. This is not surprising because a two-port with parameters {AC} represents the ideal model of the desired wideband amplifier: $Z_{IN}=A/C$, $A_{VF}=1/A$, $Z_{OUT}=0$ and $A_{VR}=0$ (figure 2.7). In this respect, two-ports {{ABC}, {ACD}, {ABCD}} are just approximation of {AC}.

Value of the non-zero {A, B, C, D} parameters. Constraints on the value of {A, B, C, D} parameters are found from the gain and stability requirements. Using equations (2.1), the gain A_{VF} of a two-port circuit with load impedance $Z_L=1/(j\omega C_L)$ can be written as:

$$A_{VF}=\frac{Z_L}{Z_L\cdot A+B} \quad \Rightarrow \quad |A_{VF}|=\frac{1}{|A+j\omega C_L B|}\leq\frac{1}{|A|}, \forall\omega \tag{2.2}$$

From (2.2), $|A_{VF}|>1$ requires a transmission parameter “A” such that $|A|<1$ holds.

Case	Z_{IN}	A_{VF}	Z_{OUT}	A_{VR}	USEFUL
{A}, {B},{AB}	∞	-	-	-	NO
{C}, {D},{CD}	0	-	-	-	NO
{AC}	A/C	1/A	0	0	YES
{AD}	$Z_L(A/D)$	1/A	-	-	NO
{BC}	$(B/C)/Z_L$	Z_L/B	-	-	NO
{BD}	B/D	Z_L/B	-	-	NO
{ABC}	$(A/C)+(B/C)/Z_L$	$1/(A+B/Z_L)$	$B/(Z_SC+A)$	$-Z_SC$	YES
{ABD}	$(B/D)+Z_L\cdot(A/D)$	$1/(A+B/Z_L)$	-	-	NO
{ACD}	$(A/C)/(1+D/(Z_LC))$	1/A	$Z_S/(Z_SC+A)$	A	YES
{BCD}	$B/(Z_L\cdot C+D)$	Z_L/B	-	-	NO
{ABCD}	$(A+B/Z_L)/(C+D/Z_L)$	$1/(A+B/Z_L)$	$(Z_SD+B)/(Z_SC+A)$	$(AD-BC)/(D+B/Z_S)$	YES

Table 2.1: Two-port transfer functions for different combinations of non-zero {A, B, C, D} transmission parameters (notation {BD} refers to two-ports parameters {0, B, 0, D}). Expressions that are not useful to the selection process are indicated by '-'.

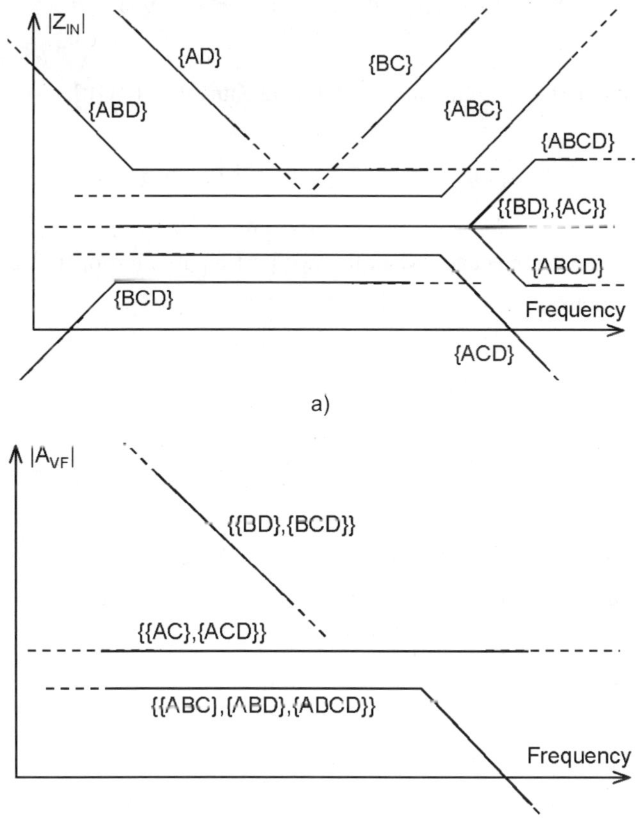

Figure 2.6: Qualitative behaviour of Z_{IN} a) and A_{VF} b) versus frequency for $Z_L=1/(j\omega C_L)$ and for different combinations of non-zero {A, B, C, D} parameters.

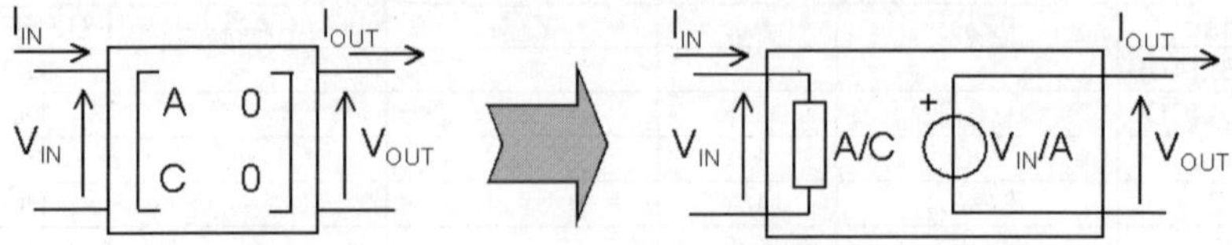

Figure 2.7: A two-port circuit with parameters {A, 0, C, 0}.

In RF and microwave amplifier design, it is a common practice to require the two-port stability to be unconditional [9]. The latter, means that the amplifier is stable for any value of the passive source and load terminations. Necessary and sufficient conditions for the unconditional stability of a linear two-port circuit are:

$$\Re\{Z_{IN}\} > 0 \quad \text{and} \quad \Re\{Z_{OUT}\} > 0 \qquad \forall\, Z_L \quad \forall\, \omega \tag{2.3-a}$$

Where $\Re$ is the real part of $\{\cdot\}$. Conditions (2.3a) are equivalent to [10]:

$$\Re\{Z_{IN}\} > 0 \quad \text{and} \quad \Re\{Z_{22}\} > 0 \qquad \forall\, Z_L \quad \forall\, \omega \tag{2.3-b}$$

Where Z_{22} is the output impedance when the two-port input is left open. Relations (2.3b) can be rewritten in terms of {A, B, C, D} parameters as:

$$\begin{aligned} \Re\{Z_{IN}\} &= \Re\left\{\frac{Z_L A + B}{Z_L C + D}\right\} > 0 \\ \Re\{Z_{22}\} &= \frac{D}{C} > 0 \qquad \forall Z_L \quad \forall \omega \end{aligned} \tag{2.4}$$

It can be shown (see appendix A) that necessary and sufficient conditions to meet relations (2.4) are that *all the {A, B, C, D} parameters must share the same sign.*

We observe that the unconditional stability requirement constraints the product of the forward gain and the reverse gain, $|A_{VF}A_{VR}|$. The latter can be written as:

$$|A_{VF}A_{VR}| = \frac{1}{|A + j\omega C_L B|} \cdot \left| \frac{AD - BC}{D + \dfrac{B}{R_S}} \right| \le \left| \frac{D - \dfrac{BC}{A}}{D + \dfrac{B}{R_S}} \right| \qquad \forall \omega \tag{2.5}$$

For $Z_{IN} \approx A/C = R_S$ and knowing that the {A, B, C, D} parameters must have the same sign, equation (2.5) yields the following inequality:

$$|A_{VF}A_{VR}| \leq \left| \frac{D - \frac{B}{R_S}}{D + \frac{B}{R_S}} \right| \leq 1 \quad \forall \omega \tag{2.6}$$

Equation (2.6) says that the product of the forward and reverse gain of an unconditionally stable matched-input two-port amplifier is lower or equal than one. In practice, a condition more stringent than (2.6) may be desired because:

- The amplifier can be considered unilateral, which means better stability [9] and lower leakage of the local oscillator signal to the amplifier input.
- The sensitivity of the input impedance to variations of the load is lower.

An important remark is that the derived constraints on the {A, B, C, D} parameters were obtained without referring to the specific nature of the two-port circuit. This means that they identify wide-band amplifiers made by any other proper set of generating elements.

2.3.3 STEP3: 2VCCS graphs database exploration

In this section, graphs of wideband amplifiers are extracted from the 2VCCS database according to the previously defined constraints on the {A, B, C, D} parameters. The table in figure 2.3b provides all the combinations of non-zero {A, B, C, D} parameters that can be realized as 2VCCS two-port circuits. However, we are interested in graphs of 2VCCS circuits according to the allowed combinations and values of non-zero transmission parameters. This selection process is outlined in table 2.2. Starting from an initial set of 145 2VCCS graphs (see appendix D), only 19 of these correspond to graphs of two-port cases: 3 {ABC}, 9 {ACD} and 7 {ABCD}. Notice that no graphs of two-port with parameters {AC} are available in the 2VCCS database. This presumably means that more than 2 VCCSs are needed to realise their functionality. The 19 graphs are then checked to verify if their {A, B, C, D} parameters can fulfil the gain and stability requirements. This possibility depends on the expression of the {A, B, C, D} parameters as a function of the transconductances g_a and g_b of the 2 VCCSs as indicated in table 2.2. For instance, all the 9 graphs of {ACD} two-ports have A=1 (i.e. they provide no gain), so they are rejected. Among the remaining 10 graphs (i.e. 3 {ABC} and 7 {ABCD}), only *1 {ABC} and 3*

{ABCD} graphs ultimately meet all the requirements. The latter are *all* the graphs of wide-band two-port amplifiers in the 2VCCS database and they are shown in figure 2.8. In the next subparagraphs, their transistor level implementations will be discussed.

2.3.4 STEP4: Transistor circuits implementation

The transistor level implementation of the graphs of 2VCCS wideband amplifiers shown in figure 2.8 depends on the orientation of the "V" and "I" branch of the VCCS and their mutual interconnection [3, 4, and 5]. Figures 2.9 shows this dependence when the V" and "I" branch share the same orientation (i.e. both arrows point to or from the same connection node). A graph with no connection between its "V" and "I" branch corresponds to a general 4-terminal VCCS element with separate input and output ports (i.e. nodes 1, 2, 3, and 4 are not connected). The latter can be implemented with a MOSFET differential pair (e.g.: n-type, p-type or complementary) or any 4-terminal transconductor circuit. If one connection exists between the "V" and "I" branch, a 3-terminal VCCS can be used. This can be implemented by a single MOSFET (either n-type or p-type depending on the arrow) or again with any 4 terminal transconductor with one of its terminals connected to one other (i.e. node 2 connected to 4). If the "V" and "I" branch are connected to each other at both ends, the 2-terminals VCCS can be implemented with a single resistor or a so-called diode-connected MOSFET. The orientation of the "V" and "I" branch also impacts the transistor implementation. For instance, reversing the orientation of both the "V" and "I" branch of a VCCS with 3 nodes its "g" is not changed while the transistor circuit changes from n-type to p-type or vice versa.

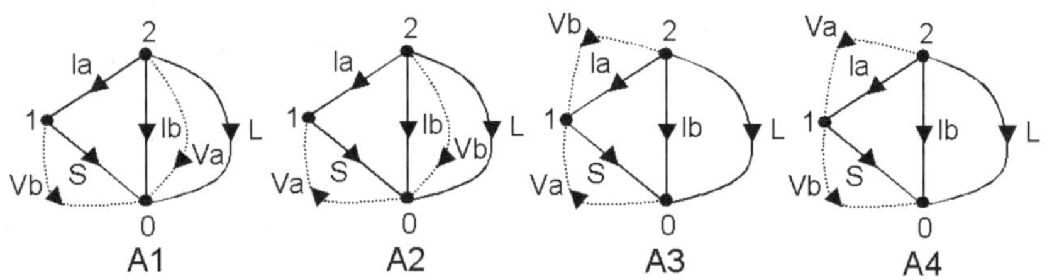

Figure 2.8: *All* the graphs of 2VCCS wideband amplifiers: A1-A4 (The symbol S, I, L over the continuous line indicates the branch of the input voltage source, the output current source of the VCCS and the load impedance Z_L respectively while the black arrows indicates the direction of the current). They are all based on the same KCL graph S+I+(I//L) described in [2,3,4] (i.e. + indicates the series connection between two branches while // a parallel one), with their "V" (i.e. the input voltage of the VCCS) branches connected to different pair of nodes (node 0: reference).

Useful Cases	Database		Graph	Transmission Parameters				|A|<1 & T.P same sign?	Condition
	Cases	Nr.		A	B	C	D		
{AC}	NO	-	-	-	-	-	-	-	-
{ABC}	{ABC}	3	**S+I+(I//L), [$V_a=V_2$, $V_b=V_1$]**	**-ga/gb**	**-1/gb**	**-ga**	**0**	**Y**	**gb>ga**
			S+I+(I//L), [$V_a=V_2$, $V_b=V_2-V_1$]	1+ga/gb	1/gb	-ga	0	N	-
			(S//I)(I//L), [$V_a=V_2$, $V_b=V_2-V_1$, $S_{ref}=0$]	1	1/gb	ga	0	N	-
{ACD}	{ACD}	9	S//(I+I)//L, [$V_a=V_2$, $V_b=V_2-V_1$]	1	0	-gagb/(ga+gb)	1	N	-
			S//(I+I)//L, [$V_a=V_2-V_1$, $V_b=V_2$]	1	0	gagb/(ga+gb)	1	N	-
			S//I//I//L, [$V_a=V_1$, $V_b=V_1$]	1	0	(ga+gb)	1	N	-
			S//(I+I)//L, [$V_a=V_1$, $V_b=V_2$]	1	0	-ga	1	N	-
			S//(I+I)//L, [$V_a=V_1$, $V_b=V_2-V_1$]	1	0	-ga	1	N	-
			(S//I//L)(I), [$V_a=V_1$, $V_b=V_2$]	1	0	ga	1	N	-
			(S//I//L)(I), [$V_a=V_1$, $V_b=V_2-V_1$]	1	0	ga	1	N	-
			(S//I//L)(I), [$V_a=V_2$, $V_b=V_2-V_1$]	1	0	ga	1	N	-
			(S//I//L)(I), [$V_a=V_2-V_1$, $V_b=V_2$]	1	0	-ga	1	N	-
{ABCD}	{ABCD}	7	**S+I+(I//L), [$V_a=-V_1$, $V_b=V_2$]**	**gb/ga**	**1/ga**	**gb**	**1**	**Y**	**ga>gb**
			S+I+(I//L), [$V_a=-V_1$, $V_b=V_2-V_1$]	**gb/(gb+ga)**	**1/(gb+ga)**	**gagb/(gb+ga)**	**ga/(gb+ga)**	**Y**	**-**
			S+I+(I//L), [$V_a=V_2-V_1$, $V_b=V_1$]	**-ga/(gb-ga)**	**-1/(gb-ga)**	**-gagb/(gb-ga)**	**-ga/(gb-ga)**	**Y**	**|gb-ga|>ga**
			S+I+(I//L), [$V_a=V_2-V_1$, $V_b=V_2$]	1+gb/ga	1/ga	gb	1	N	-
			S//I//(I+L), [$V_a=V_1$, $V_b=V_2-V_1$]	1	1/gb	ga	1+ga/gb	N	-
			S//I//(I+L), [$V_a=V_2$, $V_b=V_2-V_1$]	1	1/gb	ga	1	N	-
			(S//I)(I//L), [$V_a=V_1$, $V_b=V_2-V_1$, $S_{ref}=0$]	1	1/gb	ga	ga/gb	N	-

Table 2.2: Selection of *all* the graphs of wideband 2VCCS amplifiers (T.P. are {A, B, C, D}).

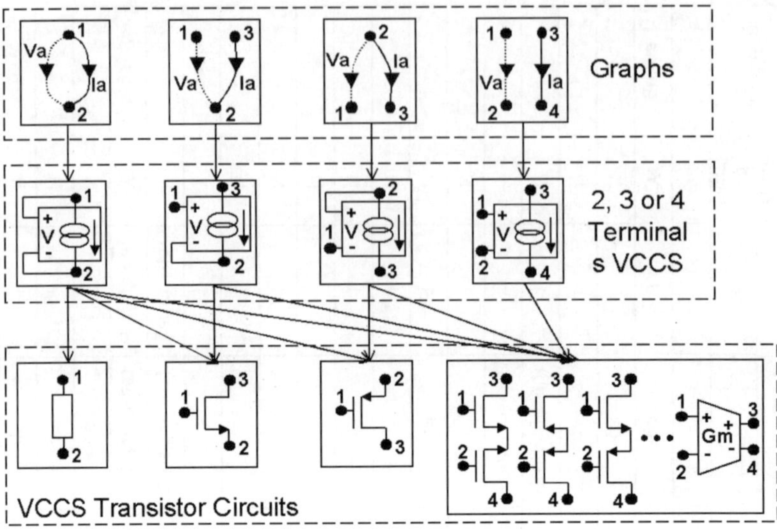

Figure 2.9: Relations among graph, VCCS and MOST implementation (biasing not shown).

On the other hand, if the "V" or "I" branch is reversed, "g" is negative. In such case, a complex circuit is required even for a 2-terminal VCCS. However, this case does not occur for the graphs in figure 2.8. In figure 2.10, the graphs of the generated 2VCCS wide-band amplifiers are draw as 2VCCSs circuits. From this figure, the conclusions are:

- Since all the VCCSs have one or even two terminals in common between the "V" and "I" port, they can be implemented using a single MOSFET (either a n-type or p-type MOSFET biased in saturation) or with a resistor (i.e. 2-terminal VCCS). These elementary amplifier implementations are shown in figure 2.11, where a generalized symbol for the MOSFET with two back-to-back arrows is used in order to cover both the n-type and the p-type MOSFET options. This means that each topology in figure 10 has 4 possible single-transistor circuit implementations: N-N, P-P, N-P and P-M where P indicate PMOS and N indicate NMOS. This corresponds to 16 different circuits. From figure 2.11, it can be seen that amplifiers A2 and A4 are well-known circuits: the common-gate and common-source shunt-feedback amplifier stages [6, 7]. The fact that well-known circuits are found using this generation methodology is not a surprise because *all* the wideband two-port amplifiers that can be modelled as circuits with 2 VCCSs have been generated. In contrast, A1 and A3 are alternative wideband amplifiers, which to the best of our knowledge are new. In [11], a circuit topology resembling A1 is proposed as a low-voltage V-I and I-I converter. However, the role of the input and output nodes was swapped, with the input signal injected into node 2 instead of node 1 (see figure 2.11).

- According to the guidelines in figure 2.9, the 2VCCS amplifiers in figure 2.10 can be implemented at transistor level in a number of different ways, each of them with specific strong and weak points. This provides more design freedom, thus potentially enhancing the quality of the design. For instance, figure 2.12 shows the amplifiers A1 and A3 where a differential pair (either n-type or p-type) replaces each VCCS. An advantage of these circuits is the absence of 2nd order distortion due to the odd-symmetry of the V-I transfer of the differential-pair.

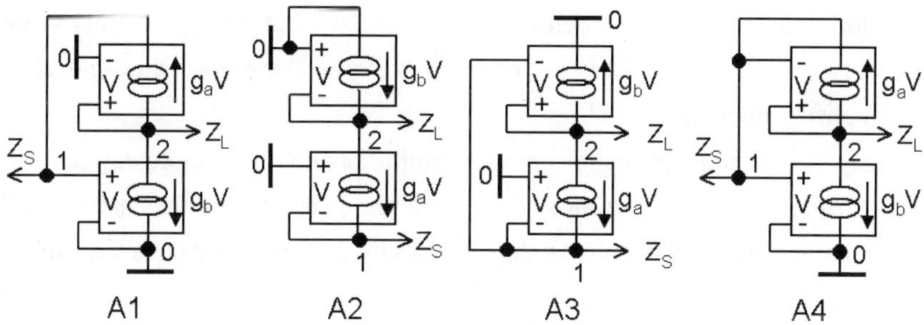

Figure 2.10: Implementations of A1-A4 using VCCS elements (biasing not shown).

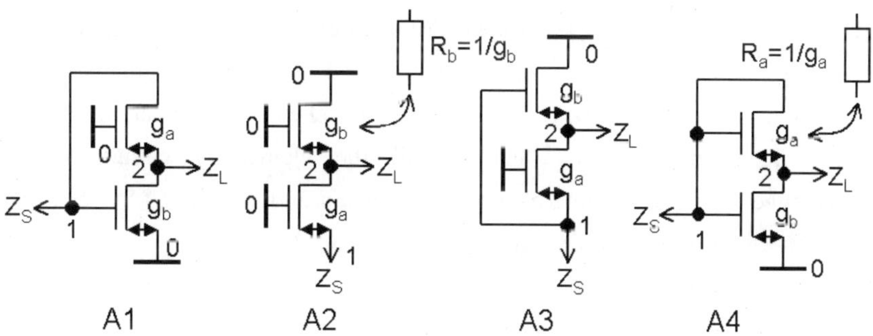

Figure 2.11: Implementation of A1-A4 with MOSFETs or resistors (biasing not shown).

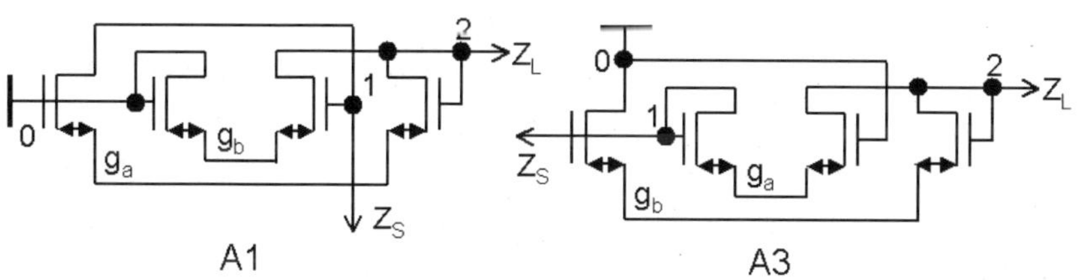

Figure 2.12: Amplifiers A1 and A3 with MOS differential pairs (biasing not shown).

2.4 Conclusions

In this chapter, alternatives to elementary wide-band amplifiers were investigated using a methodology that generates *all* the graphs of wide-band two-port amplifiers that can be seen as circuits with 2 VCCSs. This is done selecting from a previously generated graph database [3, 4, and 5] all the graphs of 2VCCS circuits that behave as wide-band amplifiers according to properly defined *functional requirements* and source/load impedance suitable for highly integrated receivers. This yielded 4 graphs of 2VCCS wideband amplifiers (figure 2.8). Examining the 2VCCS circuits in figure 2.10, it was found that:

- Replacing each VCCS element with a single MOSFET renders four elementary 2-Transistor amplifiers. Two are well-known circuits (A2 and A4, figure 2.11) while the others (A1 and A3, figure 2.11) are believed to be -at the present time- novel wide-band amplifier topologies.
- Alternative circuit implementations of these amplifiers involving more devices are also possible, each with specific strong and weak points. This provides the designer with an increased number of design options, thus potentially improving the design quality.

2.5 References

[1] Y. Tsividis, "Operation and Modelling of the MOS Transistor", Mc-Graw Hill, 2001.

[2] B. Razavi, "Design of Analog Integrated CMOS Circuits", Mc-Graw Hill, 2001.

[3] E.A.M. Klumperink, "Transconductance based CMOS Circuits: Generation, Classification and Analysis", PhD, Univ. of Twente, Enschede, The Netherlands, 1997.

[4] E.A.M. Klumperink et al, "Finding all the elementary circuits exploiting transconductance", Proc. IEEE International Symposium on Circuits and Systems, Vol. I, pp. 667-670, 6-9 may 2001, Sydney, Australia.

[5] E.A.M. Klumperink et al, "Finding all the elementary circuits exploiting transconductance", IEEE Transaction on Circuits and Systems, CAS-II, vol. 48, pp. 1039-1053, Nov. 2001.

[6] B. Razavi, "RF Microelectronic", Prentice Hall, 1998.

[7] T. H. Lee, "The Design of Radio-Frequency CMOS Integrated Circuits", Cambridge University Press, 1998.

[8] Q. Huang et al., "GSM Transceiver Front-End Circuits in 0.25-um CMOS", IEEE Journal of Solid State Circuits, pp.292-302, March 1999.

[9] G. Gonzalez, "Microwave Transistor Amplifiers", Prentice Hall, 2nd edition, 1984.

[10] M. Ohtomo, “Proviso on the Unconditional Stability Criteria of Linear Two-ports” IEEE Trans. on Microwaves Theory and Techniques, vol. 43, pp. 1197-1200, May 1995.
[11] W. Sansen et al., ”Toward Sub 1V Analog Integrated Circuits in Sub-micron Standard technologies”, Proceeding ISSCC, pp.68-71, San Francisco, USA, 1998.

Chapter 3
2-MOST Amplifiers: Analysis and Design

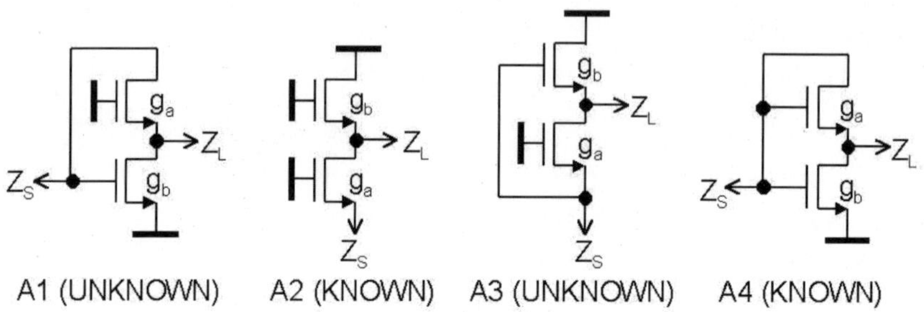

Figure 3.1: The 2-MOSFET wideband amplifiers generated systematically in chapter 2.

3.1 Introduction

In chapter 2, *all* the wideband amplifiers that can be modelled as two-port circuits with 2 VCCSs were generated systematically, yielding the 4 circuits in figure 2.10. Replacing each VCCS by a single n-type MOST, 2 well known (A2 and A4) and 2 novels (A1 and A3) wideband amplifiers were found, see also figure 3.1. These elementary amplifiers offer some advantages compared to implementations using complex VCCS circuit, as:

- A single MOSFET has a minimum number of nodes in the signal path so the amplifiers in figure 3.1 are more suitable for high frequencies.
- They are power efficient. For instance, the g_m/I_D of a common-source MOSFET is 2 or 4 times larger than for a differential pair. For equal g_m, the differential pair uses 2 or 4 times more power.
- They use a minimum of biasing sources. This leads to less parasitic capacitance and noise from the biasing circuitry is coupled into the signal path.

In this chapter, important aspects of the performance of these 2-MOSFET amplifiers are analysed. The aim is to find out whether the new amplifiers A1 and A3 perform superiorly with respect to amplifiers A2 and A4. It will turn out that amplifier A3 has favourable properties with respect to its noise factor. This result will be confirmed by the design of a 50-900MHz variable gain LNA in 0.35μm standard CMOS.

3.2 Modelling for Hand Calculations

The aim of this paragraph is to establish the model equations that will be used for a first-order comparison of the small-signal and noise performance of the amplifiers in figure 3.1. To this purpose, a MOST is modelled as a linear VCCS, $I=g \cdot V$, with "$g=g_m$". By doing so, the MOSFET output conductance g_d, body transconductance g_{mb}, terminal capacitances C_{xy} as well as the load will be neglected. This is acceptable to estimate the in-band performance of these amplifiers, as:

- *By construction*, their node impedance is determined by g_m with $g_m > \max\{g_{mb}, g_d\}$.
- *By design*, a relatively large g_m will be required to operate at high frequencies and for low-noise as well, thus relation $g_m > \omega \cdot \max\{C_{xy}, C_L\}$ holds.

For purpose of noise calculations, thermal noise associated to the conducting channel of the MOS is assumed to dominate. Its noise power spectral density is [1]:

$$\frac{\overline{I_n^2}}{\Delta f} = 4kT \cdot \gamma \cdot g_{d0} = 4kT \cdot NEF \cdot g_m \tag{3.1}$$

Where g_{d0} is the channel conductance for $V_{DS}=0$ and γ is a bias-dependent parameter. For a long-channel MOS in saturation, $\gamma=2/3$ and $g_{d0}=g_m$ holds. For a deep sub-micron MOSFET, $\gamma>2/3$ and $g_{d0}>g_m$ arise from the large electric field along the channel. Typical values of γ are between 1 and 2 [2]. To simplify the analysis, equation (3.1) is rewritten in term of the gate-transconductance of the MOST, thereby allowing a straightforward evaluation of the Signal to Noise Ratio (SNR). The noise excess factor $NEF=\gamma \cdot (g_{d0}/g_m) > \gamma$ is then introduced to quantify the excess of noise current with respect to a resistor $R=1/g_m$. This model is valid also for a resistor R if $g_m=1/R$ and $NEF=1$.

3.3 Two-port Noise Factor F

In this paragraph, the definition of noise factor F is reviewed, which will be used to compare the amplifiers of figure 3.1. The noise factor F of the two-ports in figure 3.2a driven by a signal source V_S with internal resistance R_S is defined, as [7, 11]:

$$F \equiv \frac{SNR_{IN}}{SNR_{OUT}} \quad \text{and} \quad SNR_{IN} = \frac{V_S^2}{4kT_S R_S \Delta f} \tag{3.2}$$

SNR_{IN} and SNR_{OUT} are the Signal-to-Noise Ratio at the input and output of the two-port (expressed in dB, it is often referred as noise figure $NF=10\log_{10}(F)$). F is a measure of the degradation of the SNR, which arises from noise within the two-port. For a noiseless two-port, SNR_{OUT}=SNR_{IN} and so F is 1 or NF=0dB, otherwise F is >1.

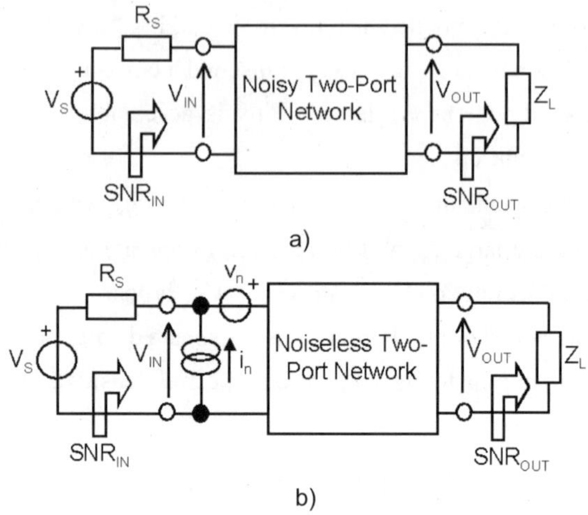

Figure 3.2: A noisy two-port driven by a resistive source (a) and its noise model (b).

Equation (3.2) is often rewritten as:

$$F = \frac{\text{Total Output Noise Power}}{\text{Output Noise Power due to the Source}}$$

In the above formulation, F is the ratio between the total noise power at the two-port output and the output noise power due only to the source. The output noise power is obtained integrating the spectral density over a range of frequencies that is relevant for the specific application. Alternatively, the noise power in 1Hz bandwidth can be used. In this case, F is often called spot noise factor. The use of one or the other definition is an application-dependent matter. For amplifiers where the signal lays in a relatively small bandwidth around a high carrier frequency, spot-F is the proper figure of merit (e.g.: RF L-C tuned LNAs). On the other hand, when the frequency-dependence of the noise power spectral densities cannot be neglected over the desired bandwidth, then a noise factor definition based on the integrated noise power is the proper measure. This is for instance the case for the front-end (e.g.: LNA+MIXER) and the base-band amplifier of a zero-IF or low-IF receiver where 1/f noise of MOSFETs is presents. In this case, the *average* noise factor F_{Avg} of the front-end can be expressed in terms of the spot F as follows:

$$F_{Avg} \equiv \frac{\int_{f_l}^{f_l+BW} A_{VF,TOT}^2(f)\cdot F(f)df}{\int_{f_l}^{f_l+BW} A_{VF,TOT}^2(f)df} = \frac{1}{BW}\int_{f_l}^{f_l+BW} F(f)df = F(f)\cdot\left[1+\frac{f_{1/f}}{BW}\cdot\ln\left(\frac{f_l+BW}{f_l}\right)\right]$$

With $[f_l, f_l+BW]$ the signal bandwidth, $f_{1/f}$ the corner-frequency of the 1/f noise at the output of the front-end and $A_{VF,TOT}$ the gain from the source V_S to the output, which is assumed frequency-independent. In the rest of this book, the spot noise figure will be used.

EQUIVALENT DEFINITIONS OF F
$F = 1 + \dfrac{\overline{V_{n,OUT}^2}}{\overline{V_{n,RS}^2}\cdot\dfrac{R_{IN}^2}{(R_{IN}+R_S)^2}\cdot A_{VF}^2}$
$F = 1 + \dfrac{\overline{V_{n,EQ,IN}^2}}{\overline{V_{n,RS}^2}}$ with $\overline{V_{n,EQ,IN}^2} \equiv \dfrac{\overline{V_{n,OUT}^2}}{\dfrac{R_{IN}^2}{(R_{IN}+R_S)^2}\cdot A_{VF}^2}$
$F = 1 + \dfrac{R_{n,EQ,IN}}{R_S}$ with $R_{n,EQ,IN} \equiv \dfrac{\overline{V_{n,EQ,IN}^2}}{4kT_S\Delta f}$
$F = 1 + \dfrac{T_{n,EQ,IN}}{T_S}$ with $T_{n,EQ,IN} \equiv \dfrac{\overline{V_{n,EQ,IN}^2}}{4kR_S\Delta f}$
$F = 1 + \dfrac{\overline{(i_n\cdot R_S + v_n)^2}}{\overline{V_{n,RS}^2}}$

Table 3.1: Alternative definitions of the noise factor F

Table 3.1 shows equivalent definitions of F in terms of equivalent input noise voltage $V_{n,EQ,IN}$, noise resistance $R_{n,EQ,IN}$, noise temperature $T_{n,EQ,IN}$ or by means of two generally correlated equivalent input noise sources i_n and v_n according to the two-port noise model in figure 3.2b. Using definition (3.2) and table 3.1 one can note that:

- F is a ratio independent on the value of the input signal.
- F depends on the value of two important parameters of the input source: the noise temperature T_S and resistance R_S. Therefore, F is meaningless and the F of different two-ports cannot be compared if T_S and R_S are not specified. To resolve this ambiguity the standard measurement procedure assumes T_S=290K, while R_S=50Ω is customary at

RF frequencies. Note as F calculated at a generic temperature T_S can be obtained from F_{290} using $F=T_{290}\cdot(F_{290}-1)/T_S+1$ because the output noise due to the two-port is not affected by the temperature T_S of the source. However, an analogous procedure to find F at a generic R_S from F@R_S=50Ω does *not* lead to a generally correct result because the two-port output noise depends on the value of R_S. This dependence is clearly shown in table 3.1, looking at the expression of F in terms of the equivalent input noise sources i_n and v_n. As R_S varies, so does the two-port output noise because of the term $i_n\cdot R_S$. When $i_n\cdot R_S<<v_n$ holds, then F at a generic R_S can be derived from F at R_S=50Ω. At RF frequencies, this is usually not the case.

- F is independent on the value of the *noiseless* load impedance Z_L. The reason is that Z_L affects the signal as well as the noise transfers of the two-port in the same way[V].

3.4 Amplifiers Performance Analysis

We now compare the performance of the amplifiers in figure 3.1. As mentioned before, amplifiers A2 and A4 are rather common circuits while amplifiers A1 and A3 are not. However, a close look to these amplifiers reveals that A1 and A3 are related to A4 and A2 respectively. In fact, amplifier A1 can be derived from A4 by connecting the gate of M_a to node "0", while amplifier A3 can be derived from A2 by connecting the gate of M_b to the input node "1". An important question now arises: Does this different interconnection of the gate terminal lead to circuits that also provide a better performance? In this section, we answer this question by looking at differences between their Z_{IN}, Z_{OUT}, A_{VF}, A_{VR} and F.

3.4.1 Small-signal transfers: Z_{IN}, Z_{OUT}, A_{VF} and A_{VR}

Table 3.2 shows Z_{IN}, Z_{OUT}, A_{VF} and A_{VR} as a function of the transconductance g_a and g_b.

AMPL.	Z_{IN}	Z_{OUT}	A_{VF}	A_{VR}
A1	$1/g_b$	$(1/g_a)/(1+g_bR_S)$	$-g_b/g_a$	g_aR_S
A2	$1/g_a$	$1/g_b$	g_a/g_b	0
A3	$1/g_a$	$1/g_b$	$1+g_a/g_b$	0
A4	$1/g_b$	$(R_S+1/g_a)/(1+g_bR_S)$	$1-g_b/g_a$	$g_aR_S/(g_aR_S+1)$
For $Z_{IN}=R_S$				
A1	R_S	$1/(2g_a)$	$-1/(g_aR_S)$	$-1/A_{VF}$
A2	R_S	$1/g_b$	$1/(g_bR_S)$	0*
A3	R_S	$1/g_b$	$1+1/(g_bR_S)$	0*
A4	R_S	$(R_S+1/g_a)/2$	$1-1/(g_aR_S)$	$1/(2-A_{VF})$

Table 3.2: Transfers of the 2-MOSFET amplifiers in figure 3.1

From this table, we observe that:

- **A1 vs. A4.** Both amplifiers exploit negative feedback to generate a well-defined input impedance $Z_{IN}=1/g_b$. For the same g_a and g_b, the forward gain A_{VF} of A4 is somewhat lower. This is because for amplifier A4 a feed-forward path via M_a does exist (i.e. the diode connected Ma acts as a *bilateral* resistor), which superimposes the input signal V_{IN} to $-(g_b/g_a)\cdot V_{IN}$ at the output node. In contrast, this feed-forward path is prevented in A1 because the gate of M_a is connected to node "0". The *unilateral* nature of M_a (i.e. $g_d=0$) hampers that the input signal V_{IN} propagates to the output through its drain. Furthermore, due to bilateral behaviour of M_a, Z_{OUT} of A4 is also somewhat larger because R_S loads its output node. From another point of view, the larger Z_{OUT} of A4 must be increased in order for both amplifiers A1 and A4 to provide equal gain A_{VF} for $Z_{IN}=1/g_b=R_S$. These results in a lower bandwidth for A4, provided A1 and A4 have the same output capacitance. The reverse gain A_{VR} of A1 is $-1/A_{VF}$ for $Z_{IN}=R_S$, while A_{VR} of A4 is $1/(2-A_{VF})$.
- **A3 vs. A2.** Both amplifiers provide well-defined input impedance $Z_{IN}=1/g_a$. Amplifier A3 is characterized by two signal paths connected in feed-forward to the output node: a path via the common gate stage M_a-M_b and the other via the source follower M_b. The latter path renders an extra +1 gain contribution for A3 provided both amplifiers use the same g_a and g_b. This means that in order for both amplifiers to provide equal gain A_{VF} for $Z_{IN}=1/g_b=R_S$, the output impedance of A2 is larger. Therefore, A3 is expected to exhibit somewhat higher bandwidth provided both amplifiers have the same output capacitance. Finally, both A3 and A2 provide good isolation as $A_{VR}=0$[VI].

3.4.2 Noise factor

Table 3.3 shows the expressions of the amplifiers F as a function of the gain $A_{VF}= V_{OUT}/V_{IN}$ for $Z_{IN}=R_S$. From table 3.3 and the NF plots in figure 3.3, we observe that:

- For all amplifiers, F is limited by the term 1+NEF, e.g.: NF≈4dB for NEF=1.5. This limitation is a direct consequence of the matching requirement $Z_{IN}=R_S$ because upon $Z_{IN}=R_S$, the matching device to generate as much equivalent input noise as that of a resistance equal to $R_S\cdot NEF$.

[V] This assumes that the load is linear time-invariant.
[VI] The input device g_d leads to $A_{VR}\neq 0$ for A3 and A2. At high frequency, A_{VR} of A3 rises due to C_{gs} of M_b.

- For all amplifiers, except A3, F increases as the gain A_{VF} drops to zero. This behaviour is rather common because the output noise power spectral density of the output device drops as $1/A_{VF}$, while the signal power at the output drops as $1/A_{VF}^2$.

AMPLIFIER	F (for $Z_{IN}=R_S$)
A1	$1+NEF_b-4NEF_a/A_{VF}$
A2	$1+NEF_a+4NEF_b/A_{VF}$
A3	$1+NEF_a+4\Delta NEF(A_{VF}-1)/A_{VF}^2$ and $\Delta NEF=NEF_b-NEF_a$
A4	$1+NEF_b(2-A_{VF})^2/A_{VF}^2+4NEF_a(1-A_{VF})/A_{VF}^2$

Table 3.3: F of the 2-MOSFET amplifiers in figure 3.1 for $Z_{IN}=R_S$.

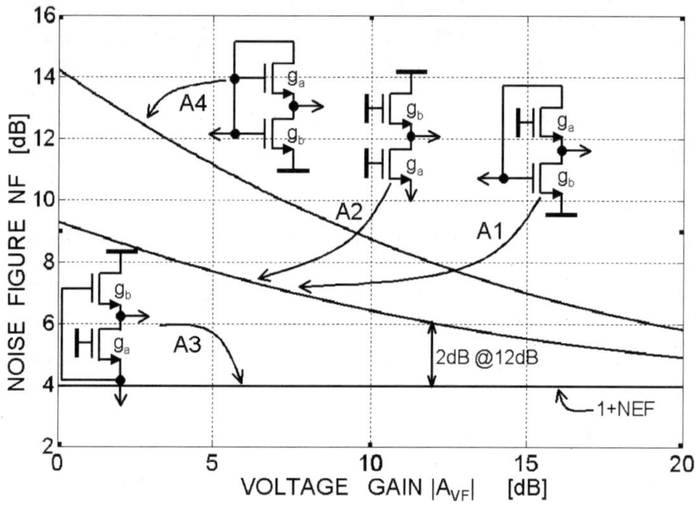

Figure 3.3: The noise figure NF versus the gain $|A_{VF}|$ for $Z_{IN}=R_S$ and $NEF_a= NEF_b=NEF=1.5$ ($\Delta NEF= NEF_a-NEF_b=0$).

- There are differences about the way amplifiers A1, A2, A3 and A4 approach the limit 1+NEF. For instance, the F of A1 is lower than that of A4 for the same gain A_{VF} and $Z_{IN}=R_S$. This can be explained in the following way. For A1 and A4, the output noise power due to the output device is equal to the noise power of an equivalent resistor NEF/g_a. However, for a given A_{VF}, A4 requires a value of g_a that is lower than for A1 (i.e. $A_{VF}=1-1/(g_aR_S)$ for A4 and $A_{VF}=-1/(g_aR_S)$ for A1). Therefore, the noise power generated by the output device is larger for A4, so F is larger too. An analogous conclusion holds when the output noise power due to the matching device is considered. In this case, half of the noise current of the input device flows into a resistor equal to R_S+1/g_a for A4 and equal to $1/g_a$ for A1. For a given A_{VF}, amplifier A4 requires a smaller "g" and so the output noise is larger. More interesting is the noise behaviour of amplifier A3. Its F is the lowest among the amplifiers in figure 3.1 and, it is *constant* as a function of A_{VF}. Specifically, NF of A3 is more than 2dB lower than that of A2 (and A3) up to A_{VF}=12dB, see figure 3.3.

To understand the reason of such behaviour, the noise factor of amplifier A3 is now analysed in detail. The noise power spectral density at the output of amplifier A3 can be separated into the contribution of the matching device M_a and the output device M_b as:

$$\frac{\overline{V_{n,OUT}^2}}{\Delta f} = \frac{\overline{V_{n,OUT,b}^2}}{\Delta f} + \frac{\overline{V_{n,OUT,a}^2}}{\Delta f} = \frac{1}{g_b^2} \cdot \frac{\overline{I_{n,b}^2}}{\Delta f} + \frac{\overline{I_{n,a}^2}}{\Delta f} \cdot \frac{(R_S - \frac{1}{g_b})^2}{(g_a R_S + 1)^2} \tag{3.3}$$

From equation (3.3), for $g_b \cdot R_S$=1 (i.e. A_{VF}=2) the output noise due to the matching device M_a is zero regardless the value of g_a (i.e. the quality of the source impedance match). This suggests some kind of *noise cancellation*.

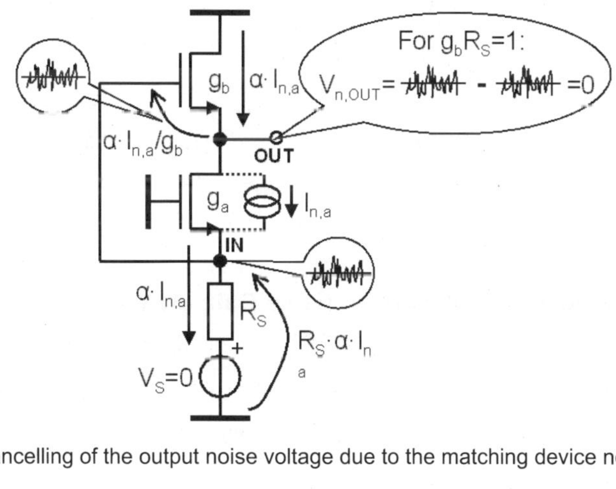

Figure 3.4: Cancelling of the output noise voltage due to the matching device noise current, $I_{n,a}$.

Figure 3.4 shows the path of the noise current outing from the input device M_a. Depending on the relation between the input impedance $Z_{IN}=1/g_a$ and R_S, a noise current $\alpha(R_S,g_a)\cdot I_{n,a}= I_{n,a}/(1+g_a\cdot R_S)$ flows out from M_a. This current leads to two *fully correlated* noise voltages $R_S\cdot\alpha(R_S,g_a)\cdot I_{n,a}$ and $\alpha(R_S,g_a)\cdot I_{n,a}/g_b$ respectively across R_S and the gate-source terminals of M_b as shown in figure 3.4. Since the output noise voltage is the instantaneous difference between the previous noise voltages, $V_{n,OUT,a}=\alpha(R_S,g_a)\cdot I_{n,a}\cdot(R_S-1/g_b)$, full output *noise cancellation* occurs for $g_b\cdot R_S$=1. In such a case, F =1+NEF_b is determined by the noise of the load device M_b. Moreover, the voltage gain A_{VF} is only 2 or 6dB. The possibility to cancel the noise current of a MOST was first discovered in a previous work of Klumperink for a different circuit [4]. For $g_bR_S\neq1$, full cancellation is prevented, so one would expect F to rise. For $NEF_a = NEF_b = NEF$ and $Z_{IN}=R_S$ equation (3.3) can be written as:

$$\frac{\overline{V^2_{n,OUT}}}{\Delta f} = \frac{NEF}{4}\cdot\frac{\overline{V^2_{n,RS}}}{\Delta f}\cdot\left[\left(1-\frac{1}{g_bR_S}\right)^2+\frac{4}{g_bR_S}\right] = \frac{NEF}{4}\cdot\frac{\overline{V^2_{n,RS}}}{\Delta f}\cdot A^2_{VF} \quad (3.4)$$

Equation (3.4) reveals that the amplifier output noise power is proportional to the product of the noise power of the source resistor R_S time the *square* of the total gain $A_{VF}/2$. Recalling the first definition of F of table 3.1, one finds the following expression regardless the value of g_b:

$$F = 1+\frac{\frac{NEF}{4}\cdot\frac{\overline{V^2_{n,RS}}}{\Delta f}\cdot A^2_{VF}}{\frac{A^2_{VF}}{4}\cdot\frac{\overline{V^2_{n,RS}}}{\Delta f}} = 1+NEF \quad (3.5)$$

This yields the *gain-independent* F shown in figure 3.3. Note that relation (3.5) strictly holds if conditions $NEF_a=NEF_b=NEF$ and $Z_{IN}=1/g_a=R_S$ are fulfilled. In the next paragraph, a wide-band LNA exploiting the noise properties of amplifier A3 is presented.

3.5 Design Example: a 50-900MHz Variable Gain LNA

In this paragraph, the design of a wideband LNA intended for wideband applications such as TV cable modem is described [5, 6]. In this respect, the following specs are assigned:

- Bandwidth: e.g.: 50-900MHz driving an on-chip capacitive load C_{EXT}=0.3pF.
- Input impedance Z_{IN}: 75Ω, with $VSWR_{IN}<2$
- Forward gain A_{VF}: 12dB (max value).
- Gain control: 4 discrete steps of 2dBs
- Noise Figure NF: <6dB.
- Reverse gain A_{VR}: ≤-20dB.
- IIP3> +2dBm and IIP2> +22dBm [10]

Wide-band amplifiers with tuneable gain are required in order to increase the front-end dynamic range and relax linearity and power requirements of the following frequency-

mixer stage[VII]. Furthermore, when the level of interference rises for a given relatively weak desired signal, it is desirable to maintain a relatively low noise figure while decreasing the gain. This is not possible using for amplifiers A1, A2 and A4 in figure 3.1 due to the noise contribution of the output device. In contrast, A3 offers the lowest $NF=10\log_{10}(1+NEF)$, which is also gain independent. Therefore, it was selected for this design.
The complete schematic of a wideband LNA based on A3 is shown in figure 3.5a.

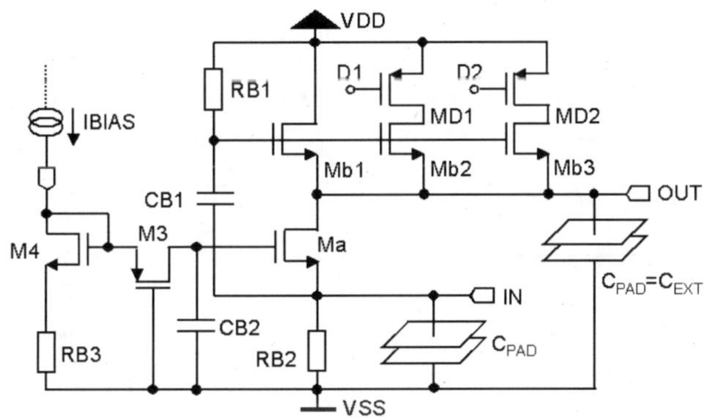

Figure 3.5: Wide-band LNA based on A3 (bond pad is used as load).

Controlled gain A_{VF} is obtained varying the width of the upper load device M_b. This is done, partitioning M_b into 3 devices M_{b1}, M_{b2} and M_{b3} with their gate, source and bulk terminals connected together, respectively. The drains of M_{b2} and M_{b3} are connected to the supply voltage via the p-type MOSFET control switches M_{D1} and M_{D2}, which set the gain. When switches M_{D1} and M_{D2} are open, the small width of M_{b1} provides the maximum gain. Conversely, minimum gain is obtained when all the switches are "on" and the effective width of the parallel of M_{b1}, M_{b2} and M_{b3} is about equal to that of M_a. Note that this position of the switches in series to the drain terminals prevents noise from the on-resistance to be coupled into the signal path. The gate terminal of M_{b1}, M_{b2} and M_{b3} cannot be dc-coupled to the source of M_a. This problem is solved using ac-coupling via the high pass filter C_{B1}-R_{B1} with a cut-off frequency $f_c=1/(2\cdot\pi\cdot C_{B1}\cdot R_{B1})<<50MHz$. At "dc", the combination of M_4 and M_a and resistors R_{B2} and R_{B3} operate as a current mirror with a large gain. The amplifier supply current is then fixed by a small external current I_{BIAS}. To guarantee the correct small-signal operation of the amplifier, capacitor C_{B2} shunts the gate

[VII] When the mixer linearity is the limiting factor, the front-end DR is maximized when the mixer is driven by an input signal maximising its S(N+D)R.

of M_a to V_{SS}. Moreover, resistor M_3 further isolates the gate of M_a from undesired signals coupled into the biasing port. This biasing scheme is preferred to a simple solution with a current mirror because resistor R_{B1} can be rather linear, has a little parasitic capacitance and it is free from 1/f noise. Finally, C_{B1} and C_{B2} are implemented by MOST capacitors to save chip-area. In the next sub-paragraphs, the amplifier behaviour is analysed.

3.5.1 Bandwidth

Figure 3.6a shows the capacitances limiting the bandwidth of the amplifier: C_1, C_L and C_2. Table 3.4 shows the expressions of C_1, C_2, C_L and the transconductance g_a and g_b.

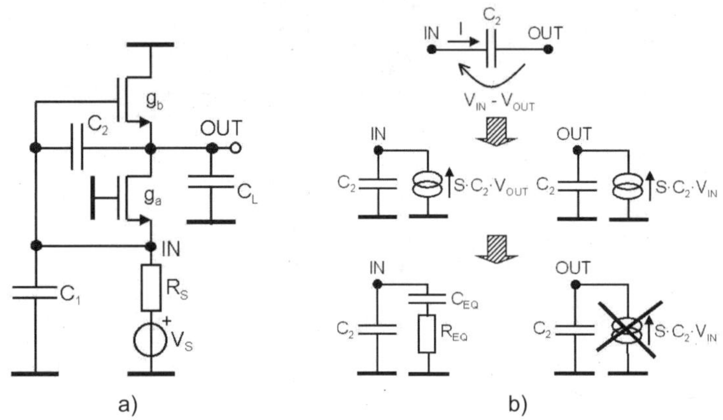

Figure 3.6: Amplifier model for high frequencies (a) and unilateral representation of C_2 (b).

C_1	C_2	C_L	g_a	g_b
$C_{gs,a}+C_{sb,a}+2C_{sb,B1}+C_{PAD}$	$C_{gs,b}$	$C_{EXT}+C_{gd,a}+C_{sb,b}+C_{db,a}$	$g_{m,a}+g_{mb,a}$	$\Sigma_i(g_{m,b,i}+g_{mb,b,i})$

Table 3.4: Expressions of the amplifier capacitance and transconductance

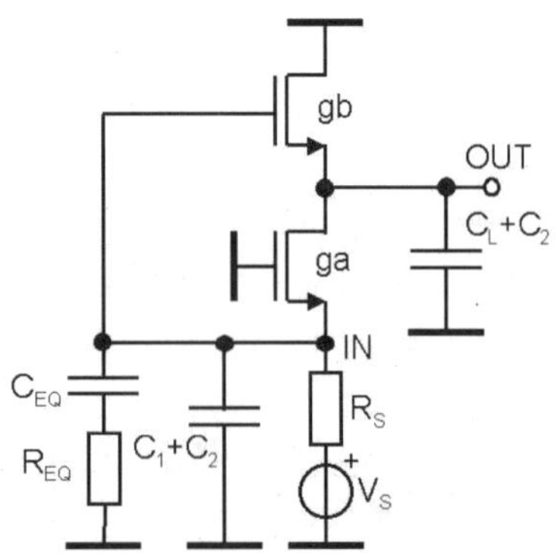

Figure 3.7: Unilateral model of the amplifier in figure 3.6a.

The *unilateral* model of C_2 is shown in figure 3.6b. Substituting this model figure 3.6a, results in the circuit of figure 3.7 with no floating capacitors. This model is valid for $\omega << g_b(1+g_a/g_b)/C_2 \approx \omega_{T,b}A_{VF}$, where $\omega_{T,b}=2\pi f_{T,b}$ is the unity current-gain cut-off frequency of Mb. Since the peak f_T of a 0.35μm MOST is in the order of 25GHz and A_{VF} is at least 2, the previous relation is satisfied up to a few GHz. Expressions for R_{EQ} and C_{EQ} are:

$$C_{EQ} = -C_2\left(1+\frac{g_a}{g_b}\right) = -C_2 A_{VF} \quad \text{and} \quad R_{EQ} = -\frac{1+\dfrac{C_L}{C_2}}{g_a+g_b} \tag{3.6}$$

R_{EQ} and C_{EQ} assume *negative* values because the gain A_{VF} is *positive*. The bandwidth, BW, is now estimated using the method of the open time-constant [7]. For $Z_{IN}=1/g_a=R_S$, the following expression for BW is found:

$$BW \approx \frac{1}{\sum \tau_i} = \frac{1}{\dfrac{C_2+C_L}{g_b} + R_S\dfrac{C_2+C_1}{2} + C_{EQ}\left(R_{EQ}+\dfrac{R_S}{2}\right)} \tag{3.7}$$

Equation (3.7) can be rewritten as:

$$BW \approx \frac{1}{\dfrac{C_2+C_L}{g_b} + R_S\dfrac{C_2+C_1+C_{EQ}}{2} + C_{EQ}R_{EQ}} \tag{3.8}$$

Substituting equations (3.6) into (3.8), one obtains:

$$BW \approx \frac{1}{2\dfrac{C_2+C_L}{g_b} + R_S\dfrac{C_1-C_2(A_{VF}-1)}{2}} \tag{3.9}$$

Table 3.5 shows the relation between $C_2=C_{gs,b}$ and $C_{gs,a}$ for different values of A_{VF}.

A_{VF}	$C_{gs,b}$	$C_{gs,b}(A_{VF}-1)$
$2 \Rightarrow W_b \approx W_a$	$\approx C_{gs,a}$	$\approx C_{gs,a}$
$3 \Rightarrow W_b \approx W_a/2$	$\approx C_{gs,a}/2$	
$N \Rightarrow W_b \approx W_a/(N-1)$	$\approx C_{gs,a}/(N-1)$	

Table 3.5: Capacitance values for different gains.

According to table 3.5, equation (3.9) can be rewritten as:

$$BW \approx \frac{1}{2\frac{C_2 + C_L}{g_b} + R_S\frac{C_{1,1}}{2}} \tag{3.10}$$

Where $C_{1,1}$ is equal to C_1 minus $C_{gs,a}$. Equations (3.9) and (3.10) show that the negative C_{EQ} pushes the input pole to higher frequencies, thereby increasing the bandwidth. For high gain (i.e. $A_{VF}>2$ or $1/g_b>R_S$), the first term of (3.10) mainly determines the bandwidth because C_2+C_L is larger than $C_{1,1}/4$. At lower gain, both terms in (3.10) are relevant. Equation (3.10) shows that the bandwidth drops as the gain increases because $C_{1,1}$ and C_L are roughly constant (i.e. the constant part of C_L and C_{EXT} dominates) and g_d drops.

The negative R_{EQ} raises some concerns about stability. According to figure 3.6a, the input admittance, $Y_{IN}(s)$, is equal to:

$$Y_{IN}(s) = g_a + s\cdot(C_1 + C_2) + \frac{s\cdot C_{EQ}}{1 + s\cdot R_{EQ}\cdot C_{EQ}} \tag{3.11}$$

$$Y_{IN}(j\omega) = g_a + \frac{\omega^2\cdot C_{EQ}^2\cdot R_{EQ}}{1+\omega^2\cdot C_{EQ}^2\cdot R_{EQ}^2} + j\omega\cdot\left(C_1 + C_2 - \frac{C_{EQ}}{1+\omega^2\cdot C_{EQ}^2\cdot R_{EQ}^2}\right) \tag{3.12}$$

The real part of $Y_{IN}(\omega)$ is equal to:

$$\Re\{Y_{IN}(\omega)\} = g_a + \frac{\omega^2\cdot C_{EQ}^2\cdot R_{EQ}}{1+\omega^2\cdot C_{EQ}^2\cdot R_{EQ}^2} \tag{3.13}$$

A sufficient condition for stability is: $\Re\{Y_{IN}(\omega)\}\geq 0\ \forall\omega$. This leads to the condition:

$$g_a \geq \max_\omega\left\{-\frac{\omega^2\cdot C_{EQ}^2\cdot R_{EQ}}{1+\omega^2\cdot C_{EQ}^2\cdot R_{EQ}^2}\right\} = -\frac{1}{R_{EQ}} = \frac{g_a + g_b}{1+\frac{C_L}{C_2}} \tag{3.14}$$

For $g_a=1/R_S$, $g_b\leq 1/R_S$ and $C_L/C_2\geq 1$, equation (3.14) is satisfied. In practice, stability can be even better than what is showed by equation (3.14) because:

- The conductance $1/R_{B2}$ (see figure 3.6a) adds to the first term of equation (3.14).

- Equation (3.14) holds for $\omega \rightarrow \infty$. At these frequencies, an oscillation is unlikely to start.
- A negative real part of the input impedance leads to instability only if larger than $-R_S$.

3.5.2 Noise factor

The actual value of the noise factor is affected by a number of noise sources, which were not considered in the simplified analysis carried out previously. In the following, a more complete analysis of the noise factor is provided. This is done by splitting the analysis between low-medium and high operation frequencies.

For frequencies well below the –3dB bandwidth, the F of the amplifier in figure 3.5 is:

$$F_{LNA} = F + (F_{EXTRA} - 1) \tag{3.15}$$

The noise factor F is obtained from table 3.3; while F_{EXTRA} accounts for noise sources that have not been considered in F. Contributions to F_{EXTRA} arise mainly from:

- The distributed gate resistance of M_a and M_b and that of the substrate beneath.
- The thermal noise from the biasing devices.
- The 1/f noise of M_a and M_b.

However, some of these noise sources can be made negligibly small. For instance, the distributed gate resistance of the MOST is small when the number of gate-fingers is large enough and the gate terminal contacted at both sides [4]. The resistance associated to the substrate beneath the MOSFETs can be significantly decreased using a large number of substrate contacts around the device and in between the gate fingers [7]. Proper layout for the input pad is also necessary in order to prevent noise from the substrate resistance being coupled to the amplifier input and output ports via the bond pads [8]. On the other hand, noise from the biasing devices and 1/f noise have a less negligible impact at these frequencies. Assuming $Z_{IN}=R_S$, $R_{B2}/R_S>>1/2$, $R_{B1}/R_S>>A_{VF}/2$ and $f_c<<50$MHz, F_{EXTRA} is:

$$F_{EXTRA} - 1 \approx \frac{R_S}{R_{B2}} + \frac{4 \cdot R_{B1}}{A_{VF}^2 \cdot R_S} \cdot \frac{1}{1 + \omega^2 \cdot R_{B1}^2 \cdot C_{B1}^2} + (F_{1/f} - 1) \tag{3.16}$$

The first term of equation (3.16) represents the contribution of resistor R_{B2} when C_{B2} shunts the gate of M_a to V_{SS} so that the noise contribution of M4, M3 and R_{B3} is negligible. This term can be made small for $R_{B2}>>R_S$. The second and the third term of equation (3.16) are due to the resistance R_{B1} of the high-pass filter and the 1/f noise of the MOSFET in the signal path. The contribution of R_{B1} to F is small when most of its noise

voltage drops across R_{B1} itself. This is ensured by choosing R_{B1} and C_{B1} such that $R_{B1} >> |1/(j\cdot\omega\cdot C_{B1}) + R_S/2|$ holds for $f<f_c<50MHz$. A large R_{B1} OR a large C_{B1} can be used to this purpose. The term $F_{1/f}$ relates to the 1/f noise of M_a and M_b. Since minimum channel length is used to maximise speed and M_a/M_b have a low/moderate width, the output 1/f noise in the low MHz range can be substantial. In this respect, poly resistors for R_{B1} and R_{B2} instead of a MOST resistor and current source help to lower the 1/f noise.
At high frequencies, F increases because the amplifier parasitic capacitances cause the output signal power to drop faster than the noise power. To analyse this effect, the amplifier model of figure 3.6a is used, where the capacitance from the output node to ground is removed because irrelevant as far as the behaviour of F is concerned. For this circuit, the frequency-dependent noise factor for $Z_{IN}=R_S$ and $NEF_a=NEF_b=NEF$ is:

$$F(\omega) = 1 + NEF \cdot \frac{1 + \omega^2 \cdot R_S^2 \cdot (C_1^2 \cdot A_{VF} + C_2^2 + 2 \cdot C_1 \cdot C_2) \cdot \frac{A_{VF}-1}{A_{VF}^2}}{1 + \omega^2 \cdot R_S^2 \cdot C_2^2 \cdot \frac{(A_{VF}-1)^2}{A_{VF}^2}} \tag{3.17}$$

Equation (3.17) shows how C_1 and C_2 degrade F with the frequency. The effect of these capacitances can be intuitively understood using the following reasoning. Consider initially the case $C_2=0$. Capacitance C_1 decreases the gain A_{VF} as the frequency increases. However, C_1 does not affect the output noise due to M_b. Therefore, the contribution of M_b to F increases with the frequency. Let's now analyse the contribution of M_a. Part of its noise current flows into the equivalent (complex) impedance $R_S//(1/(j\cdot\omega\cdot C_1))$. As the frequency increases, the noise voltage across $R_S//(1/(j\cdot\omega\cdot C_1))$ does not subtract perfectly to the noise voltage across the gate-source terminals of M_b. In fact, this creates a frequency zero $\omega_z=1/(R_S C_1)$ in the transfer function, which increases the output noise (and so F) with the frequency. From another point of view, the presence of this zero can be seen observing that the output signal$\rightarrow 0$ for $\omega\rightarrow\infty$, while the output noise due to M_a is not (i.e. max for $\omega\rightarrow\infty$). For $C_2\neq 0$ the same conclusion holds, however, the increase of F with the frequency is lower because C_2 renders a zero at $\omega_z=-(g_a+g_b)/C_2$ in the signal transfer.

3.6 Design

The design of the amplifier in figure 3.5 is straightforward. Since Ma and Mb's conduct the same current and minimum channel length is mandatory in order to minimise parasitic capacitances, three design parameters are left: W_a, $W_{b's}$ and $V_{GS,a}-V_{T0}$. If W_a and $W_{b's}$ are fixed by the matching and gain requirements (i.e. $g_a=1/R_S$, $g_b=R_S(A_{VF}-1)$ and $g_m=g_m(W, V_{GS}-V_{T0})$), the only unknown is $V_{GS,a}-V_{T0}$. However, considerations concerning power, gain, F and linearity suggest the use of a relatively small $V_{GS,a}-V_{T0}$, because:

- For $Z_{IN}=R_S$, power dissipation $P=V_{DD}I_D=(V_{DD}/R_S)(I_D/g_a)$ drops as $V_{GS,a}-V_{T0}$ decreases.
- The gain $A_{VF}=1+(g_a/I_D)/(g_b/I_D)\approx1+(V_{GS,b}-V_{T0})/(V_{GS,a}-V_{T0})$ rises as $(V_{GS,a}-V_{T0})$ drops. Alternatively, given A_{VF} the minimum supply voltage is lower if $V_{GS,a}-V_{TO}$ is small.
- The contribution to F from the bias resistor R_{B2} is small if R_{B2} is large. This reduces the voltage headroom for the rest of the circuit. For a given gain A_{VF} and V_{DD}, the value of R_{B2} is the largest when $(V_{GS,a}-V_{T0})$ is small.
- Large $V_{GS,a}-V_{T0}$ is not necessarily needed to achieve good intercept points, IIP2 and IIP3. For $A_{VF}=2$, Ma and Mb have equal size, bias current and similar drain source voltage. In this condition, the intercept points can be high because the non-linear V-I conversion performed by Ma is cascaded to a *near-inverse* non-linear I-V conversion of Mb. For gain values larger than 6dB, the previous non-linearity compensation holds to a lesser extent and the IIPs are expected to decrease.

ITEM	VALUE
W_a	120μm
W_{b1}, W_{b2}, W_{b3}	20μm, 20μm, 40μm
RB2	750Ω
RB1	100KΩ
CB1	2pF (220μm /2μm)
I_{BIAS}	30μA
I_{DD}	1.4mA
V_{DD}	3.3Volt

Table 3.6: Sizing of the amplifier in figure 3.5

According to the previous considerations, the amplifier was designed in a standard 0.35μm CMOS process. Table 3.6 shows design parameters obtained using $V_{GS,a}-V_{T0}\approx220mV$. A $R_{B2}=750\Omega$ degrades NF of about 0.1dB compared to the case of an ideal current source, while keeping enough voltage headroom for the rest of the circuit. A large poly-silicon R_{B1} with a relatively small C_{B1} saves area and reduces the back-plate parasitic, which increases of F at high frequencies. Specifically, $R_{B1}=100K\Omega$ and $C_{B1}=2pF$ (i.e. $f_c<1MHz$) increase NF less than 0.2dB at 50MHz. Figure 3.8 shows the simulated max and min voltage gains $A_{VF}=V_{OUT}/V_{IN}$ and $A_{VF,TOT}=V_{OUT}/V_S$ versus frequency using MOS model 9. The max value of A_{VF} is about 11.5dB with a bandwidth BW of 1.03GHz. The latter is close to 0.85

GHz estimated using equation (3.10). Figure 3.9 shows the simulated real part of the input impedance versus frequency. Its value remains positive for all the frequencies. Figure 3.10 shows the simulated voltage standing wave ratio, $VSWR_{IN}$, versus frequency. Its value degrades with the frequency due to the input capacitance. Nevertheless, it remains below 2 up to 1GHz. Figure 3.11 shows the reverse gain A_{VR} versus frequency. At lower frequency, input-output isolation is limited by the output conductance of Ma. At higher frequencies, the reverse gain increases due to the feed-forward path to the input through capacitance C_2. A_{VR} remains lower than -30dB up to 1GHz.

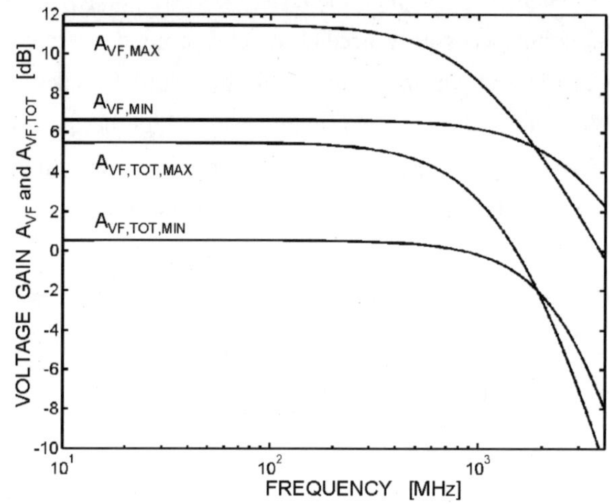

Figure 3.8: Simulated amplifier voltage gains A_{VF} and $A_{VF,TOT}$.

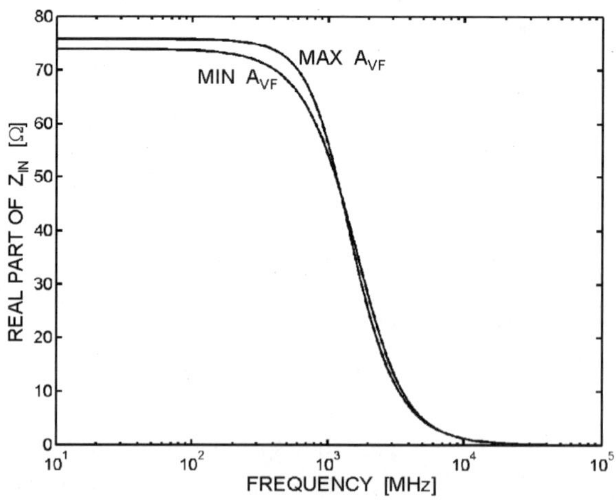

Figure 3.9: Simulated real part of the input impedance versus frequency.

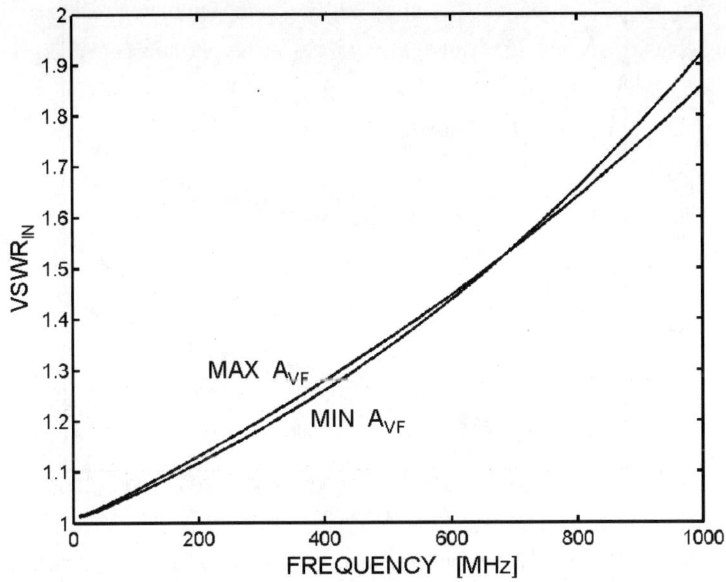

Figure 3.10: Simulated input voltage standing wave ratio versus frequency.

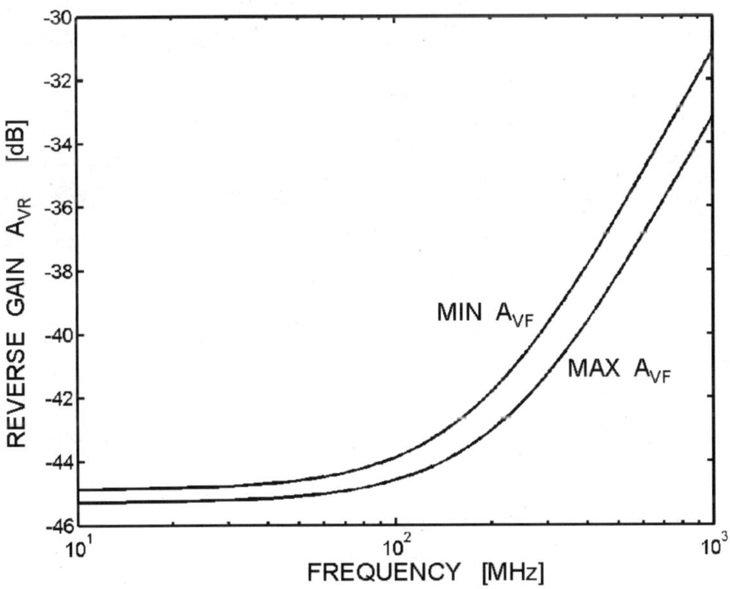

Figure 3.11: Simulated reverse gain A_{VR}=20log$_{10}$(V_{IN}/V_{OUT}) versus frequency.

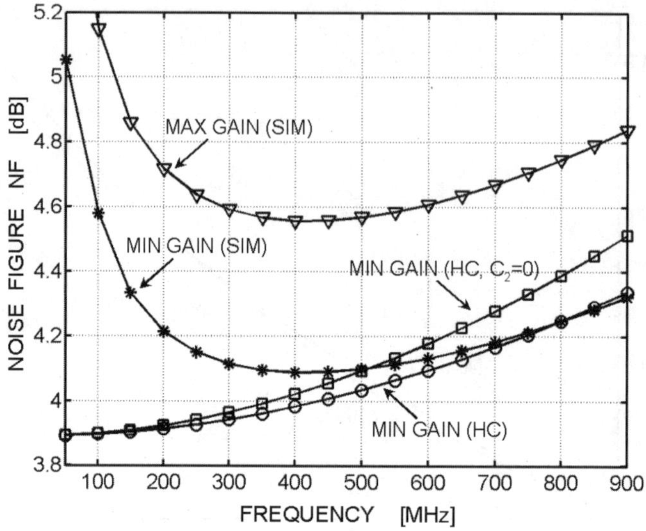

Figure 3.12: NF for R_S=75Ω versus frequency (HC=Calculations; SIM=Simulations).

Figure 3.12 shows the simulated noise figure referred to R_S=75Ω versus frequency for max and min gain. The noise figure is somewhat dependent on the gain with a maximum variation of about 0.45dB. This is because NEF is bias-dependent via γ and g_{d0}/g_m and the input matching isn't perfect. Nevertheless, the simplified analysis of F still captures the essential behaviour of the circuit. At lower frequencies, the noise figure degrades due to the small C_{B1} used in the high-pass filter. At higher frequencies, both simulation and hand calculations show a modest increase of NF. Equation (3.17) is also plotted in figure 3.12 for a min gain, showing a good agreement with the simulation. This is expected because M_a and M_b experience about the same V_{GS} and V_{DS}, so $NEF_a \approx NEF_b$ holds. The hand-calculated curve for C_2=0 is also plotted, which overestimates the increase of F.
The input-referred 2nd and 3rd order IM intercept points IIP2 and IIP3 have been simulated for two input tones located at (200MHz, 500MHz) and (450MHz, 500MHz), respectively. At minimum gain, IIP2 and IIP3 are +24dBm and +13.5dBm respectively. At maximum gain, IIP2 and IIP3 are +28.5dBm and +14dBm. These numbers are high if the gate-source voltage of Ma is not large. Furthermore, the IIPs do not change significantly with the gain. The compression performance of the amplifier, which is essentially limited by its class-A operation, is –6dBm and –4dBm at minimum and maximum gain respectively (i.e. desired signal at 200MHz). These values differ significantly from IIP3–9.6dB predicted from simple theory [7, 11].

3.7 Measurements

Figure 3.13a shows the chip-photo of the wide-band CMOS LNA of figure 3.5. The LNA transfer functions A_{VF}, A_{VR} and $VSWR_{IN}$ were obtained from its [S]-parameters measured on-wafer. To do so, the [S]-parameters measured with respect to 50Ω were converted to a reference of 75Ω using well-know two-port formulas. For example, the LNA input reflection coefficient with respect to a reference impedance R_1, $\Gamma_{IN,R1}$, is equal to:

$$\Gamma_{IN,R1} = \frac{Z_{IN} - R_1}{Z_{IN} + R_1} = S_{11,R1} + \frac{S_{21,R1} S_{12,R1} \Gamma_{L,R1}}{1 - S_{22,R1}} \tag{3.18}$$

Where $\Gamma_{L,R1}$ is the load reflection coefficient. Equation (3.18) can be used to find the input reflection coefficient with respect to a reference impedance R_2, $\Gamma_{IN,R2}$, as:

$$\Gamma_{IN,R2} = \frac{\Gamma_{IN,R1} \dfrac{R_1 + R_2}{R_1 - R_2} + 1}{\Gamma_{IN,R1} + \dfrac{R_1 + R_2}{R_1 - R_2}} \tag{3.19}$$

From equation (3.19) for R_1=50Ω and R_2=75Ω, $VSWR_{IN,75}$ can be then calculated as:

$$VSWR_{IN,75} = \frac{1 + |\Gamma_{IN,75}|}{1 - |\Gamma_{IN,75}|} \tag{3.20}$$

With $\Gamma_{L,R1}$=1 as the load capacitance C_L is absorbed into the LNA. In the following, the measured transfer functions A_{VF}, A_{VR} and $VSWR_{IN}$ are referred to 75Ω.

In figure 3.13b, the input voltage standing wave ratio $VSWR_{IN}$ is <1.6 up to 900 MHz and is marginally affected by the gain settings. Figure 3.14a and 3.14b show the forward gain and the reverse gain A_{VF} and A_{VR} versus frequency for different gain settings. A_{VF} ranges from 6.2dB to 11dB, which is somewhat lower than expected from simulations. The worst-case (i.e. at maximum gain) –3dB bandwidth is 950 MHz. This value is rather close to 0.85GHz and 1.03GHz predicted from calculations and simulations respectively. The reverse gain A_{VR} is < -30dB up to 900 MHz with a $|A_{VR}A_{VF}|$<-19 dB. According to simulations, A_{VR} increases with the frequency due to the input-output capacitance.

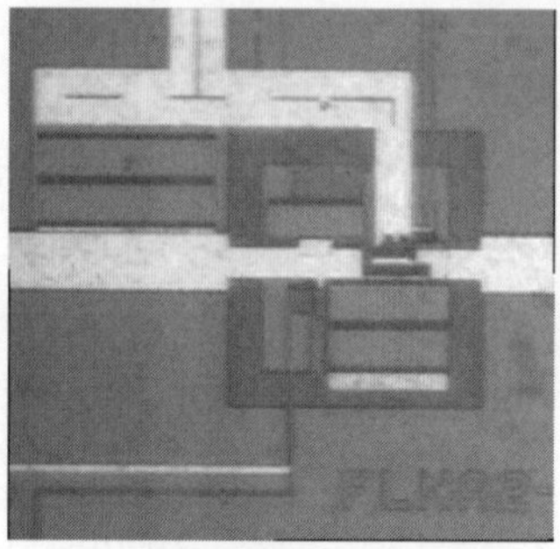

a)

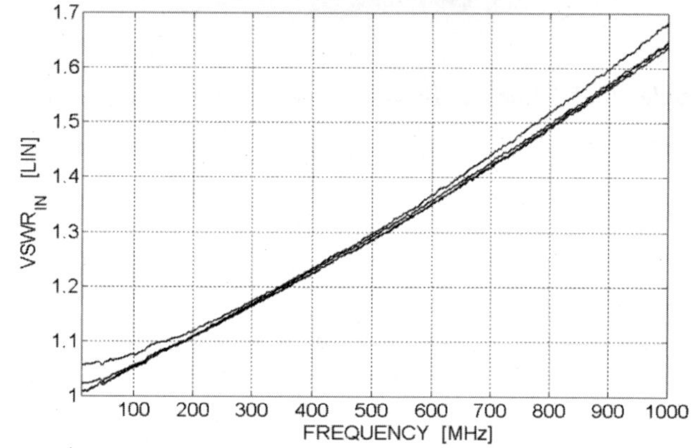

b)

Figure 3.13: LNA chip-photo (a) and measured $VSWR_{IN}$ versus frequency for different gains (b).

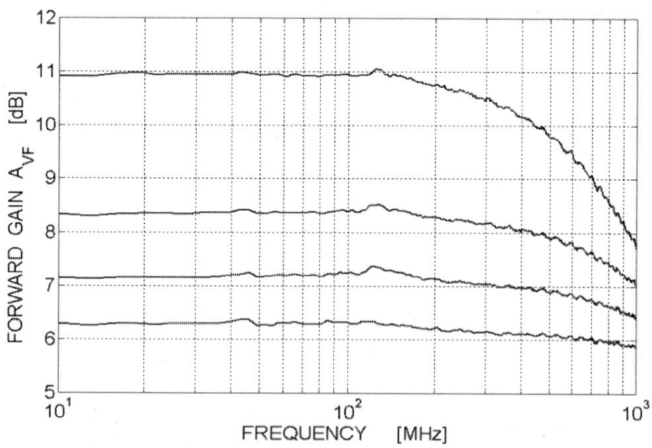

a)

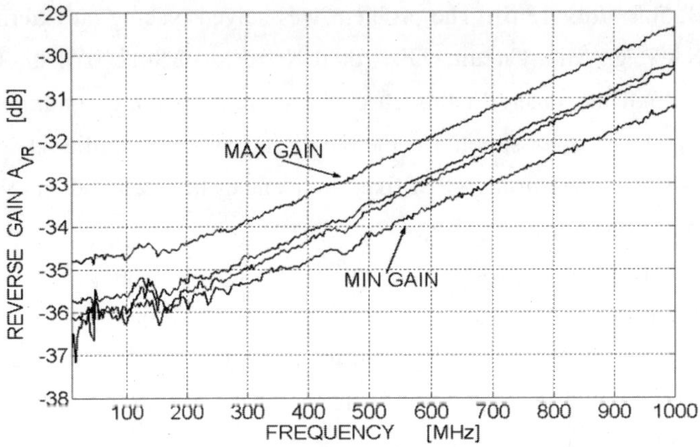

b)

Figure 3.14: Measured A_{VF} (a) and A_{VR} (b) versus frequency for different settings.

The amplifier NF was measured with the chip mounted on a PCB and connected to 75Ω transmission lines. In order to measure noise figure with a NF-meter (e.g.: HP8970B), a few problems were faced:

1. The device under test (DUT) input/output impedance must be matched to 50Ω.
2. For maximum accuracy, DUT and meter must fulfil [10]: $NF_{DUT}+G_{DUT}>5dB+NF_{METER}$ where $G_{DUT}=R_S[A_{VF}/(R_S+R_{IN})]^2/R_{OUT}$ is the *available power* gain of the DUT.
3. The above issues must be resolved over a wide range of frequencies.

The first point is of concern because the DUT input impedance is 75Ω while its output impedance can be as large as 240Ω at maximum gain! A microwave tuner was then placed behind the DUT to match its output impedance to 50Ω for each gain setting. On the other hand, it was decided not to use another tuner to match the DUT input to 50Ω, thereby measuring NF with respect to a 50Ω input source. This choice was motivated as follows:

- From simulations, the noise figure and its variations with respect to the forward voltage gain are somewhat larger (i.e. a few fraction of dB) for a 50Ω input source.
- The relatively modest input mismatch is expected to render a measurement error more than an order of magnitude smaller than the targeted NF value. Moreover, since the input mismatch is near the same for all the gain settings (i.e. $|A_{VF}A_{VR}|$<-19dB), the measurement error is also expected to be the same.

The second point is also crucial because the LNA has little available power gain, G_{DUT}. Even for $G_{DUT}=G_{TUNER}=0dB$, $NF_{DUT}=4dB$ and $NF_{METER}=7dB$, yields the impossible condition 4dB>5dB+7dB=12dB! The problem was solved placing an external high-gain wide-band LNA (e.g.: Mini-Circuits ZFL-1000LN: NF_{TYP}=2.9dB, BW=0.001-1GHz and G_{MIN}=20dB) behind the tuner. In this case, the more favourable condition 20dB+4dB >12dB is obtained where, for simplicity, the NF of the external amplifier is neglected. The third point was solved measuring NF at fixed frequencies between 400 and 900 MHz due to the limitations of the tuner.

The LNA noise factor was measured using the set-up of figure 3.15. The automatic calibration of the NF-meter was used to correct for the contribution of the blocks following the post-amplifier (i.e. cable-2 and the meter itself). After noise calibration, F_m read in the display refers to the 50Ω noise factor of the cascade of blocks between the reference planes (i.e. dashed lines).

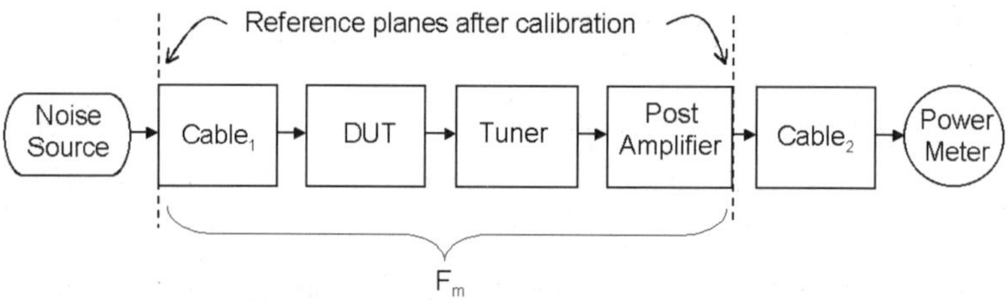

Figure 3.15: Block diagram of the set-up to measure the noise figure.

The measured noise factor F_m can be written as [11]:

$$F_m = F_{CABLE,1} + \frac{F_{DUT}-1}{G_{CABLE,1}} + \frac{F_{TUNER}-1}{G_{CABLE,1}\cdot G_{DUT}} + \frac{F_A-1}{G_{CABLE,1}\cdot G_{DUT}\cdot G_{TUNER}} \tag{3.21}$$

where each F in equation (3.21) is referred to the output impedance of the previous stage and the G's are available power gains. Equation (3.21) can be rewritten as:

$$F_m = G_{CABLE,1}\cdot F_{DUT} + \frac{F_A - G_{TUNER}}{G_{CABLE,1}\cdot G_{DUT}\cdot G_{TUNER}} \tag{3.22}$$

Where $F_{CABLE,1}=1/G_{CABLE,1}$ and $F_{TUNER}=1/G_{TUNER}$ were used. From (3.22), F_{DUT} is:

$$F_{DUT} = \left(F_m - \frac{F_A - G_{TUNER}}{G_{CABLE,1}\cdot G_{DUT}\cdot G_{TUNER}}\right)\cdot\frac{1}{G_{CABLE,1}} \tag{3.23}$$

In order to determine F_{DUT}, the noise figure F_A of the post amplifier, the available gains G_{TUNER}, $G_{CABLE,1}$ and G_{DUT} were measured separately. Following an analogous procedure, the LNA noise figure F_{LNA} was obtained from F_{DUT} after de-embedding the contribution of the PCB transmission lines.

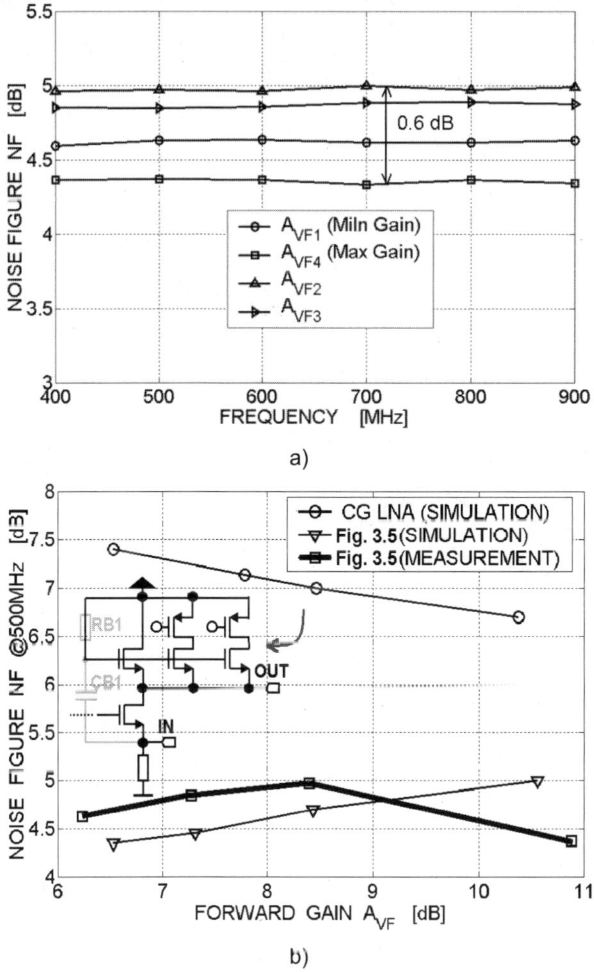

Figure 3.16: Measured NF vs. frequency for different gains (a) NF @500MHz versus A_{VF} for the LNA of figure 3.5 and the derived CG amplifier (b).

The LNA noise figure is shown in figure 3.16a versus frequency for different gains. NF@50Ω varies between 4.3dB and 4.9dB, which exceeds somewhat the max variation found the simulated NF@75Ω in figure 3.12. The measured NF@75Ω and the simulated

NF@75Ω at 500MHz are shown in figure 3.16b versus the gain A_{VF}. The simulated NF@75Ω of a common-gate (CG) amplifier is also shown. The latter has been obtained from the amplifier of figure 3.5 by connecting the gates of $M_{b1,2,3}$ to V_{DD} (i.e. removing the path of the high-pass filter C_{B1}-R_{B1}) and resizing their W's for the same gain steps. According to the expectations, the LNA in figure 3.5 provides about *constant NF at least 2dB better than that of the CG amplifier upon the same A_{VF}, Z_{IN} and power.*

Figure 3.17a shows the extrapolated IIP2 and IIP3 at maximum gain. Figure 3.17b shows IIP2, IIP3 and 1dBCP versus the gain. These are rather close to the simulation results. A summary of the measurements at maximum gain is shown in table 3.7.

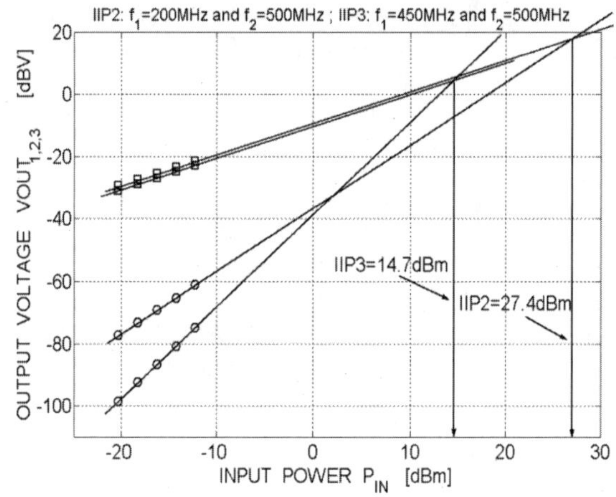

a)

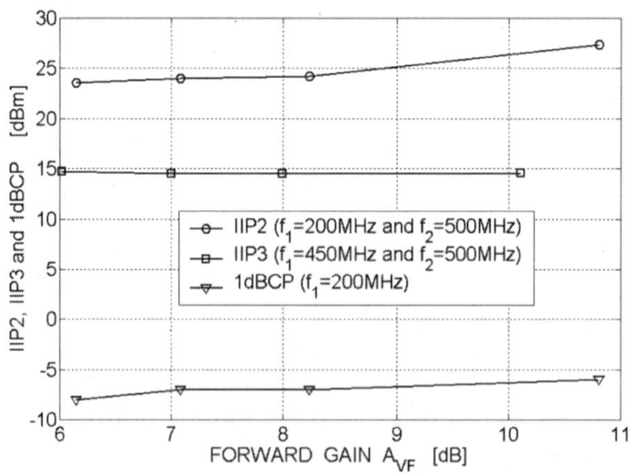

a)

Figure 3.17: Measured IIP2 and IIP3 at maximum gain (a) and IIP2, IIP3 and 1dBCP vs. gain (b).

PROPERTY	VALUE
$\lvert A_{VF}=V_{OUT}/V_{IN}\rvert$	11 dB
-3dB BW	1-900MHz (C_{EXT}=0.28pF)
$\lvert A_{VR}=V_{IN}/V_{OUT}\rvert$	< -30 dB up to 900 MHz
$VSWR_{IN}$	< 1.6 up to 900 MHz
IIP3 (input ref.)	14.7 dBm
IIP2 (input ref.)	27.4 dBm
ICP1 (input ref.)	-6 dBm
$NF_{50\Omega}$	<4.4 dB
I_{DD} @ V_{DD}	1.5mA @ 3.3Volt
Technology & Die area	0.35μm CMOS & 0.06 mm^2

Table 3.7: Summary of the LNA measurements at maximum gain

3.8 Conclusions

In this chapter, 2-MOSFET implementations of the wide-band amplifiers generated in chapter 2 were compared. Based on hand-calculations it was found that the newly found amplifier topologies A1 and A3 offer a superior noise factor and small-signal performance with respect to the well-known amplifiers A2 and A4 (see figure 3.1) because:

- For the same value of the device transconductance g_a and g_b, A1 provides a larger forward voltage gain compared to A4. Conversely, for the same A_{VF}, C_L and Z_{IN}=R_S, A1 offers a somewhat larger bandwidth compared to A4. The same conclusions hold also for A3 with respect to A2.
- For Z_{IN}=R_S and same voltage forward gain, the noise factor of A1 is lower than that of A4 due to its lower output noise power. However, A3 showed the lowest noise factor equal to 1+NEF among the considered amplifier that is constant as a function of the gain. Specifically, the NF is at least 2dB lower than that of the other amplifiers upon the same gain, Z_{IN}=R_S and power dissipation. This behaviour is due to a cancellation mechanism for the output noise due to the matching device. Because of its superior noise performance, the new amplifier A3 was chosen to design a 50-900MHz variable-gain LNA in 0.35μm standard CMOS. Prototype measurements confirmed a NF between 4.3dB and 4.9dB (versus gain) at least 2dB better than that of the other amplifiers for the same Z_{IN}, gain A_{VF} and power.

Finally the results achieved in this chapter proves that the systematic generation methodology can lead to new circuit topologies of wide-band amplifiers that also have superior performance with respect to existing solutions.

3.9 References

[1] A. van der Ziel, "Thermal Noise in Field effect transistors", Proceedings IEEE, pp. 1801-1812, August 1962.

[2] A.J. Scholten et al.,"Accurate thermal noise modelling for deep-submicron CMOS", International Electron Device Meeting IEDM, Digest of Technical papers, 1999.

[3] E.A.M. Klumperink, "Transconductance based CMOS Circuits: Generation, Classification and Analysis", PhD. Thesis, University of Twente, Enschede, The Netherlands, 1997.

[4] B. Razavi et al., "Impact of distributed gate resistance on the performance of MOS devices", IEEE Trans. on Circuits and Systems—I, vol. 41, pp. 750-754, Nov. 1994.

[5] F. Bruccoleri, E.A.M. Klumperink and B. Nauta "Generating *All* 2-MOS Transistors Circuits Leads to New Wide-Band CMOS LNAs", Proceedings of 26th European Solid-State Circuits Conference ESSCIRC' 2000, pp.288-291, Kysta, Sweden, 19-21 Sept. 2000.

[6] F. Bruccoleri, E.A.M. Klumperink and B. Nauta "Generating *All* 2-MOS Transistors Amplifiers Leads to New Wide-Band LNAs", IEEE Journal of Solid-State Circuits, vol. 36, pp. 1032-1040, July 2001.

[7] T. H. Lee, "The Design of Radio-Frequency CMOS Integrated Circuits", Cambridge University Press, 1998.

[8] A. Rofougaran, et al., "1 GHz CMOS RF front-end IC for a direct-conversion wireless receiver", IEEE Journal of Solid-State Circuits, vol.30, pp. 1399-1410, December 1995.

[9] A. van der Ziel, "Noise in Solid State Devices and Circuits", Wiley, New York 1986.

[10] H. Packard, "Noise Figure Measurement Accuracy", Application Note 57-2, Nov. 1988.

[11] B. Razavi, "RF Microelectronic", Prentice Hall, 1998.

[12] B. Taddiken, "Broadband Design for Wireless and Wired Systems", Tutorial, ISSCC 2001, San Francisco, USA.

Chapter 4
Wide-Band Low-Noise Techniques

4.1 Introduction

In chapter 2, *all* the wide-band two-port amplifiers that can be modelled as circuits with 2 voltage controlled current sources were generated systematically. In chapter 3, elementary implementations of these amplifiers using 2 MOSTs were studied. From their analysis, a noise cancellation mechanism was discovered, which lead to the design of a wide-band variable-gain amplifier with a *lower noise figure* for the same gain and power compared to known elementary circuits. In this chapter, traditional wide-band low-noise techniques suitable for monolithic integration are reviewed and their noise properties and limitations highlighted. Finally, a novel low-noise technique is proposed as alternative to the aforementioned approaches.

Along this chapter, unless otherwise stated, three assumptions will be extensively exploited to simplify the (small-signal) analysis of amplifiers, yet capturing their essential behaviour:

- The MOST is modelled as an ideal Voltage Controlled Current Source (VCCS).
- Biasing is done by means of ideal voltage and current sources, thus only the behaviour of the devices in the signal path is considered.
- Thermal noise associated to the conductive channel of the MOST or a resistor is assumed the dominant source of noise.

4.2 Noise Factor Considerations

In this paragraph, the (spot) noise factor F of a wide-band amplifier is analysed in terms of the contributions due to its internal devices. The aim is to gain insight about what device limits the noise performance of the amplifier and how this limitation does occur. To this purpose, F is regarded as the ratio between the signal-to-noise ratio (SNR) at the amplifier input and output ports. Reasoning in terms of SNR is preferred to other equivalent formulations (e.g.: see chapter 3) because it highlights a major concern in the design of a receiver: to guarantee a certain minimum SNR when handling the weakest input signal.

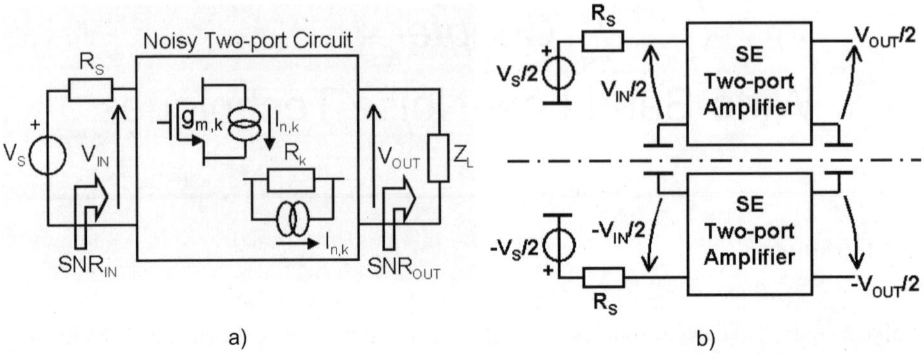

Figure 4.1: a) Two-port model of a single-ended (SE) wide-band amplifier built using MOSFETs and resistors b) Balanced amplifier using two identical separable SE two-port amplifiers.

Consider the two-port wide-band amplifier in figure 4.1a, which is built using M devices like MOSTs and resistors. Its noise factor F can be written as:

$$F \equiv \frac{SNR_{IN}}{SNR_{OUT}} = \frac{SNR_{IN}}{\dfrac{S_{OUT}}{N_{OUT,S}+\sum_{k=1}^{M} N_{OUT,k}}} \tag{4.1a}$$

Where $N_{OUT,S}$ and $N_{OUT,k}$ are noise powers at the output of the two-port due to the source and the 'k-th' device respectively. The SNR at the input of the two-port, SNR_{IN}, is equal to the SNR of the input source V_S, $SNR_S=V_S^2/(4kT \cdot R_S\Delta f)$. Equation (4.1a) can be written as:

$$F = 1+\sum_{k=1}^{M} \frac{SNR_{IN}}{SNR_{OUT,k}} \tag{4.1b}$$

where $SNR_{OUT,k}=S_{OUT}/N_{OUT,k}$ is the SNR at the output of the two-port when only the noise of the 'k-th' device is active. Equation (4.1b) yields to another expression of F:

$$F = 1+\sum_{k=1}^{M} \frac{N_{OUT,k}}{N_{OUT,S}} = 1+\sum_{k=1}^{M} \frac{N_{OUT,k}}{N_{IN,S} \cdot A_{VF,TOT}^2} \tag{4.1c}$$

where $A_{VF,TOT}=V_{OUT}/V_S$ is the total voltage gain and $N_{IN,S}$ is the noise generated by the source resistor, R_S. Both equations (4.1b) and (4.1c) can be written as follows:

$$F = 1 + \sum_{k=1}^{M} EF_k \tag{4.2a}$$

$$EF_k \equiv \frac{SNR_{IN}}{SNR_{OUT,k}} = \frac{N_{OUT,k}}{N_{IN,S} \cdot A_{VF,TOT}^2} \tag{4.2b}$$

Above, the *excess* noise factor, EF_k, is introduced in order to quantify the degradation of F due to noise generated by the 'k-th' device. From these equations, the following important considerations are derived:

- F is determined by the sum of the excess noise factor EF_k of each device. For instance, a noiseless amplifier (i.e. F=1) requires $EF_k=0$ for any k=1…M.
- For a noisy amplifier, values of F below 2 (i.e. NF<3dB) lead to the condition $EF_k<1$ $\forall$ k=1…M or equivalently:
  - $SNR_{OUT,k}>SNR_{IN}$ $\forall$ k=1…M or
  - $N_{OUT,k}<N_{OUT,S}$ $\forall$ k=1…M.

  This means that if (at least) one device "k" exists such EF_k is larger than one, then F is always larger than 2 (i.e. $F>1+EF_k$ must hold).

The same conclusions can be extended to a balanced amplifier made using two equal and *distinct* single-ended (SE) two-port amplifiers driven by two signal sources in anti-phase and equal amplitude (Figure 4.1b). Also in this case, an expression of F analogous to (4.1b), (4.1c), (4.2a) and (4.2b) can be found using the following substitutions:

$$\left\{ R_S \mapsto R_{S,D} \; ; F \mapsto F_D \; ; \; \frac{SNR_{IN}}{SNR_{OUT}} \mapsto \frac{SNR_{IN,D}}{SNR_{OUT,D}} \; ; \; \frac{N_{OUT,k}}{N_{OUT,S}} \mapsto \frac{N_{OUT,D,k}}{N_{OUT,D,S}} \; ; \; EF_k \mapsto 2 \cdot EF_{D,k} \right\}$$

where the subscript 'D' refers to a differential quantity and the factor 2 is due to the fact that the number of devices (in the signal path) is doubled. Although a balanced amplifier contains two times the number of noise sources, its F_D referred to a differential source resistance $R_{S,D}=2 \cdot R_S$ is equal to the F of its constituting SE amplifier when referred to R_S[VIII]. This is because the two times larger output noise due to the balanced amplifier is compensated by a two times larger output noise due to the source (i.e. $R_{S,D}=2 \cdot R_S$). However, the balanced LNA consumes twice as much power than its SE version.

[VIII] Noise of the biasing circuitry that appears as a common mode signal at the output of the balanced LNA increases the F of its single-ended part stand-alone. This noise contribution is considered negligible.

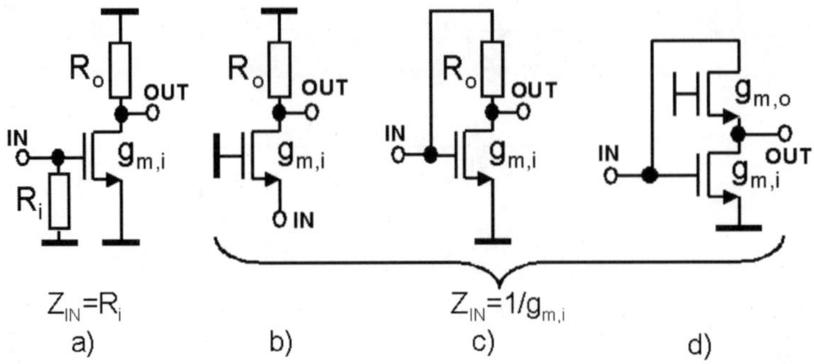

Figure 4.2: Elementary wide-band LNAs (biasing not shown): a) CS stage plus resistive termination (CST), b) CG stage, c) CS shunt feedback stage (CSSF) and d) amplifier A1 ("i" and "o" refer to the input and output devices).

4.3 F to Impedance Matching Trade-off in Elementary Wide-band LNAs

In this paragraph, limitations to the noise factor of known elementary wide-band LNAs are discussed. Figure 4.2 shows some examples of elementary amplifiers: the common source stage with resistive input termination (CST, a), the common gate stage (CG, b), the CS shunt-feedback stage (CSSF, c) and amplifier A1 presented in chapter 3 (d). According to equation (4.2), limitations to the noise factor are tied to their device EF_k. However, not all the devices affect F in the same fashion. For these amplifiers, two types of devices are distinguished. The device determining the input impedance OR the V-I conversion (i.e. input device, indicated with 'i') and the device providing the impedance transformation (i.e. output device, indicated with 'o') necessary for a voltage gain larger than one. Table 4.1 shows EF_i and EF_o as a function of $g_{m,i}R_S$, R_S/R_i and the total gain $A_{VF,TOT}= V_{OUT}/V_S$.

LNA	$A_{VF,TOT}$	EF_I	EF_o
a	$-g_{m,i}R_o/(1+R_S/R_i)$	R_S/R_i $NEF\cdot(1+R_S/R_i)^2/(g_{m,i}R_S)$	$-(1+R_S/R_i)/(g_{mi}R_SA_{VF,TOT})$
b	$g_{m,i}R_o/(1+g_{mi}R_S)$	$NEF/(g_{m,i}R_S)$	$(1+g_{m,i}R_S)/(g_{mi}R_SA_{VF,TOT})$
c	$(1-g_{m,i}R_o)/(1+g_{mi}R_S)$	$(NEF/(g_{m,i}R_S))\cdot(1-A_{VF,TOT})^2/A_{VF,TOT}^2$	$(1-(1+g_{mi}R_S)A_{VF,TOT}))/(g_{mi}R_SA_{VF,TOT}^2)$
d	$-(g_{m,i}/g_{m,o})/(1+g_{mi}R_S)$	$NEF/(g_{m,i}R_S)$	$-(1+g_{m,i}R_S)/(g_{mi}R_SA_{VF,TOT})$
Upon Source Impedance Matching: $Z_{IN}=R_S$			
a	$-R_o/(2R_S)$	1 $4\cdot NEF/(g_{m,i}R_S)$	$-2/(A_{VF,TOT}g_{mi}R_S)$
b	$R_o/(2R_S)$	NEF>1	$2/A_{VF,TOT}$
c	$(1-R_o/R_S)/2$	$NEF\cdot(1-A_{VF,TOT})^2/A_{VF,TOT}^2>NEF$	$4\cdot(1-2A_{VF,TOT})/A_{VF,TOT}^2$
d	$-1/(2g_{m,o}R_S)$	NEF>1	$-2/A_{VF,TOT}$

Table 4.1: EF_i and EF_o for the amplifiers in figure 4.2. NEF is the noise excess factor (see the definition in chapter 3) and $A_{VF,TOT}=V_{OUT}/V_S$.

The input impedance Z_{IN} is equal to $1/g_{m,i}$ for amplifiers CG, CSSF and A1 and to R_i for the CST stage. For a given gain, the following limits for EF_i and EF_o can be found:

$$\lim_{\substack{g_{mi}R_S\to\infty \\ \text{or}\ \frac{R_i}{R_S}\to\infty}} \{EF_i\} = 0 \quad \text{amplifiers 'a,b,c,d'}$$

$$\lim_{\substack{g_{mi}R_S\to\infty \\ \text{or}\ \frac{R_i}{R_S}\to\infty}} \{EF_o\} = \begin{cases} \dfrac{1}{|A_{VF,TOT}|} & \text{, amplifiers 'b,c,d'} \\ 0 & \text{, amplifier 'a'} \end{cases}$$

The above results can be explained as follows:

- **EF_i:** For amplifiers ‘b, c and d’, the noise current outing the input device is equal to zero for $g_{m,i}R_S\to\infty$ (i.e. it flows into the device itself because $Z_{IN}\to 0$). On the other hand, the signal current flowing into the output device is equal to V_S/R_S for $g_{m,i}R_S\to\infty$, so $SNR_{OUT,i}\to\infty$. For amplifier ‘a’, the output noise of the input termination R_i is zero for $R_i/R_S\to\infty$ while the signal current is $g_{m,i}V_S$, again yielding $SNR_{OUT,i}\to\infty$. For the input MOST of amplifier ‘a’, the same conclusion holds because the signal current increases with $g_{m,i}$ while the noise current is only proportional to the square root of $g_{m,i}$.
- **EF_o:** For amplifiers ‘b, c and d’, as $g_{m,i}R_S\to\infty$ the output voltage tends to $R_O\cdot V_S/R_S = |A_{VF,TOT}|\cdot V_S$ -or $V_S/(g_{m,o}R_S)=|A_{VF,TOT}|\cdot V_S$- while the output noise voltage is proportional to the square root of $|A_{VF,TOT}|$ (via R_O or $g_{m,0}$). For amplifier ‘a’, $EF_o\to 0$ for $g_{m,i}R_S\to\infty$ because the output voltage is proportional to $g_{m,i}$ while the output noise voltage is constant (for a given gain).

From the above analysis, for $|A_{VF,TOT}|>>1$ and both $g_{m,i}R_S$ and $R_i/R_S\to\infty$, F drops with the input device EF_i until its value is limited by the output device $EF_o\approx 1/|A_{VF,TOT}|<<1$ (for ‘a’, both EF_i and EF_o tend to zero). This highlights a fundamental trade-off between their F and the source impedance matching requirement $Z_{IN}=R_S$. *On one hand, low values of F require $g_{m,i}R_S$ and R_i/R_S to be much larger than one. On the other hand, source impedance matching demands a fixed $g_{m,i}R_S=1$ and $R_i/R_S=1$.* This means that, for $Z_{IN}=R_S$, the excess noise factor of the matching device is already $NEF\geq 1$ (i.e. NEF=1 for a resistor; NEF>1 for a MOST). Therefore, $F>1+EF_i>2$ (i.e. NF>3dB) holds because the matching device contributes to F at least as much as the input source does.

4.4 Working Around the Trade-off

In this paragraph, two ways to relax somewhat the trade-off between F and Z_{IN}=R_S are described: capacitive input cross coupling and source impedance mismatch.

Figure 4.3a shows the schematic of a balanced CG amplifier stage exploiting capacitive input cross coupling (CC) [2]. In contrast to the traditional CG amplifier stage (figure 4.3a, R=0 and C=0), cross coupling capacitors are used to allow the entire differential input voltage $V_{IN,D}$ to drop across the gate and source terminals of each of the input MOSTs, thereby enhancing their effective transconductance. To do so, the gate of each MOST is connected to the source of the other via a dc-level shifter (e.g.: high-pass filters in figure 4.3a). The F of the cross-coupled amplifier is analysed using the model in figure 4.3b assuming R=C=∞.

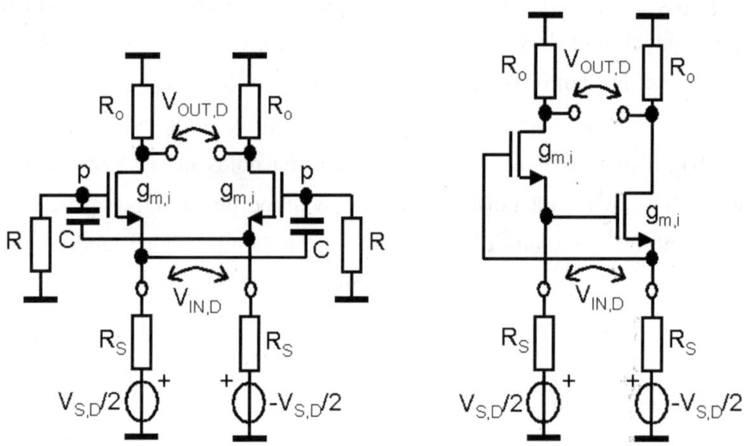

a) b)

Figure 4.3: Balanced CG amplifier with input capacitive cross coupling (biasing not shown). Simplified model used for hand calculations for R=C=∞.

Case	**$R_{IN,D}$**	**$R_{IN,C}$**	**$G_{m,D}$**	**$A_{VF,TOT,D}=V_{OUT,D}/V_{S,D}$**
NO CC	$2/g_{m,i}$	$1/g_{m,i}$	$g_{m,I}$	$G_{m,i}R_o/(1+g_{m,i}R_S)$
CC	$1/g_{m,i}$	$1/g_{m,i}+R_S$	$2 \cdot g_{m,i}$	$2 \cdot g_{m,i}R_o/(1+2 \cdot g_{m,i}R_S)$
Upon Impedance Matching: $Z_{IN,D}=R_{S,D}$ with $R_{S,D}=2 \cdot R_S$				
NO CC	$2 \cdot R_S$	R_S	$1/R_S$	$R_o/(2 \cdot R_S)$
CC	$2 \cdot R_S$	$3 \cdot R_S$	$1/R_S$	$R_o/(2 \cdot R_S)$

Table 4.2a: Small-signal properties of the balanced amplifier

Case	$2 \cdot N_{OUT,D,I}$	$2 \cdot EF_{D,i}$	$2 \cdot EF_{D,o}$
NO CC	$2 \cdot R_o^2 I_{n,i}^2/(1+g_{m,i}R_S)^2$	$NEF/(g_{m,i}R_S)$	$(1+g_{m,i}R_S)/(g_{m,i}R_S A_{VF,TOT,D})$
CC	$2 \cdot R_o^2 I_{n,i}^2/(1+2 \cdot g_{m,i}R_S)^2$	$NEF/(4 \cdot g_{m,i}R_S)$	$(1+2 \cdot g_{m,i}R_S)/(2 \cdot g_{m,i}R_S A_{VF,TOT,D})$
Upon Impedance Matching: $Z_{IN,D}=R_{S,D}$ with $R_{S,D}=2 \cdot R_S$			
NO CC	$R_o^2 I_{n,i}^2/2$	NEF	$2/A_{VF,TOT,D}$
CC	$R_o^2 I_{n,i}^2/4$	NEF/2	$2/A_{VF,TOT,D}$

Table 4.2b: Device EF_D of the amplifier in figure 4.3a (CC = Cross Coupling).

Tables 4.2a shows the differential input resistance $R_{IN,D}$, common mode resistance $R_{IN,C}$, transconductance $G_{m,D}$ and gain $A_{VF,TOT,D}$ for a differential source resistance $R_{S,D}=2 \cdot R_S$ and a differential output resistance $R_{o,D}=2 \cdot R_o$. Table 4.2b shows the devices excess noise factor $EF_{D,i}$ and $EF_{D,o}$ evaluated for the case with and without cross coupling (i.e. $R=C=\infty$ and $R=C=0$). According to these tables, $G_{m,D}$ is two times larger using cross coupling and so the input impedance $R_{IN,D}$ is also two times smaller. For $Z_{IN,D}=R_{S,D}$, the $g_{m,i}$ used by the cross coupled LNA is two times smaller, thus halving power consumption. In this case:

1. Both the LNAs provide the same differential $G_{m,D}$ and gain $A_{VF,TOT,D}$.
2. The matching devices $2 \cdot EF_{D,i}$ is NEF/2 using cross coupling because the output noise power is halved (i.e. $g_{m,i}$ is the half) while the output signal power stays the same.

The cross-coupled LNA provides then a lower F while using half of the power! However, cross coupling suffers from important limitations:

- F_D is fundamentally limited to 1+NEF/2 because the trade-off with $Z_{IN,D}=R_{S,D}$ stands still. For high-sensitivity applications this value is just not enough low.
- Antennas, cables and high-frequency filters are typically single-ended devices. This means that a single-ended to balanced conversion (via the so-called balun) must be performed prior the LNA. Such operation involves always a degradation of the SNR, so the cascade $F_{Balun+LNA}$ can be larger than F_{LNA}. Furthermore, a passive discrete wide-band balun increases manufacturing costs, occupies significant PCB area and can couple interference at the input node due to its relatively large physical size.
- The voltage drop across the cross coupling C's (caused by the capacitance from node "p" to ground C_p and the gate-source capacitance $C_{gs,CC}$) degrades $Z_{IN,D}$, F_D and gain at all frequencies (table 4.2c). Figure 4.4a shows $NF_D(\omega=0)$ vs. $C/C_{gs,CC}$ for $Z_{IN,D}=2 \cdot R_S$, $C_p=C_{gs,CC}$, NEF=1.5 and $A_{VF,TOT,D}=5$. The ratio $C/C_{gs,CC}$ must exceed 15 to degrade NF less than 0.1dB. This can require C in the order of 10pF [2].
- The matching MOSTs load each other, resulting in a larger input time constant $\tau_{IN,CC}$ (tables 4.2c and 4.2d) and so a faster degradation of F_D and gain at high frequencies.

Figure 4.4b shows the ratio $\tau_{IN,CC}/\tau_{IN}$ versus $C/C_{gs,CC}$. Even thought that $C_{gs,CC}$ is twice smaller than C_{gs} (i.e. for the same g_m/I_D), $\tau_{IN,CC}/\tau_{IN}$ is larger than 2 for $C/C_{gs,CC}=10$.

- Finally, as the signal applied to each MOST is two times larger distortion increases.

Case	**No CC**	**CC**
$Z_{IN,D}(\omega=0)$	$2/g_{m,I}$	$(1+(C_p+C_{gs,CC})/C)/[g_{m,i}(1+C_p/(2C))]$
$A_{VF,TOT,D,i}(\omega=0)$	$g_{m,i}R_o/(1+g_{m,i}R_S)$	$[2g_{m,i}R_o(1+C_p/(2C))/(1+(C_p+C_{gs,CC})/C)]/[1+2R_S/Z_{IN,D}(\omega=0)]$
$\tau_{IN}\ (Z_{IN,D}=2R_S)$	$R_SC_{gs}/2$	$[R_SC_{gs,CC}/2](4+C_p/C_{gs,CC}+C_p/C)/(1+(C_p+C_{gs,CC})/C)$

Table 4.2c: Small-signal properties of the amplifier in figure 4.3a accounting for the effect of the parasitic capacitances C_p and C_{gs}. (CC = Cross Coupling)

Case	$2\cdot EF_{D,i}(\omega)$
NO CC	$[NEF/(g_{m,i}R_S)]\cdot[1+(\omega\cdot R_SC_{gs})^2]$
CC	$[NEF/(4g_{m,i}R_S)][(1+(C_p+C_{gs,CC})/C)/(1+C_p/(2C))]^2[1+ +(\omega R_SC_{gs,CC}(4+C_p/C_{gs,CC}+C_p/C)/(1+(C_p+C_{gs,CC})/C))^2]$
Upon Impedance Matching: $Z_{IN,D}=R_{S,D}$ with $R_{S,D}=2\cdot R_S$	
NO CC	$NEF[1+(\omega R_SC_{gs})^2]$
CC	$[NEF/2][(1+(C_p+C_{gs,CC})/C)/(1+C_p/(2C))][1+ +(\omega R_SC_{gs,CC}(4+C_p/C_{gs,CC}+C_p/C)/(1+(C_p+C_{gs,CC})/C))^2]$

Table 4.2d: Input device $EF_{D,i}$ for the balanced amplifier in figure 4.3a accounting the effect of the parasitic capacitances C_p and C_{gs}.

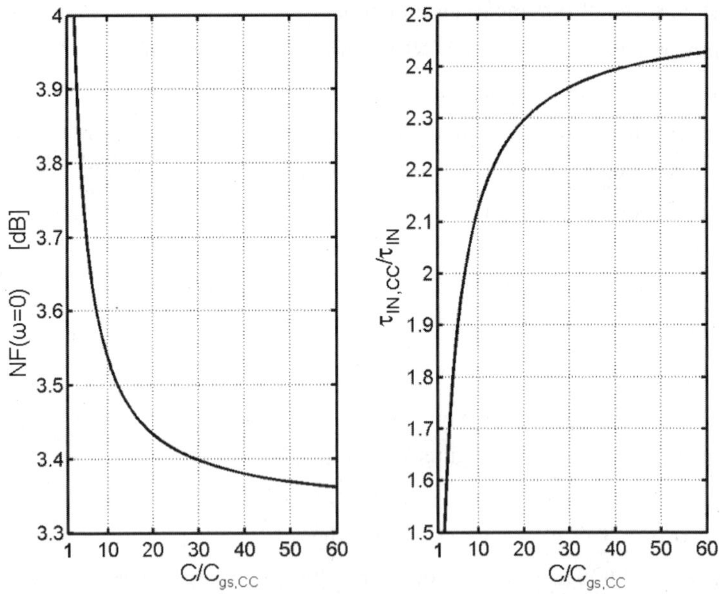

Figure 4.4: a) $NF(\omega=0)$ versus $C/C_{gs,CC}$ and (b) $\tau_{IN,CC}/\tau_{IN}$ versus $C/C_{gs,CC}$ ($NEF=1.5$, $A_{VF,TOT,D}(\omega=0)=5$ and $C_p=C_{gs,CC}$).

So far, accurate impedance matching $Z_{IN(,D)}=R_{S,(D)}$ was assumed. In practice, a certain application-dependent impedance mismatch is tolerated. For instance, the RF filter preceding the LNA in a mobile receiver tolerates terminations whose reflection coefficient, $|\Gamma_{IN(,D)}=(Z_{IN(,D)}-R_{S(,D)})/(Z_{IN(,D)}+R_{S(,D)})|$, can be as large as -10dB [4]. This enable the possibility to lower the noise factor of the LNAs in figure 4.2 and 4.3 by mismatching their input according to: $g_{m,i}R_{S(,D)}>1$ (i.e. $\Gamma_{IN}<0$) or $R_i/R_S>1$ (i.e. $-\Gamma_{IN}>0$). Table 4.3 shows their F as a function of the input reflection coefficient $\Gamma_{IN(,D)}$, the total gain $A_{VF,TOT(,D)}$ and for equal power consumption. Their NF is plotted in figure 4.5 versus the $|\Gamma_{IN}|_{dB}$ for $|A_{VF,TOT}|=5$ and NEF=1.5. Clearly, for $|\Gamma_{IN}|_{dB}$ close to -10dB, only the cross-coupled LNA provides an NF below 3dB. Nevertheless, NF is still limited by $NF(|\Gamma_{IN}|_{dB}=|\Gamma_{IN}|_{dB,MAN})$.

LNA	Noise Factor F
CST	$1+(1+\Gamma_{IN})/(1-\Gamma_{IN})+4\cdot NEF/(1-\Gamma_{IN}^2)-2/(A_{VF,TOT}\cdot(1-\Gamma_{IN}))$
CG	$1+NEF\cdot(1+\Gamma_{IN})/(1-\Gamma_{IN})+2/(A_{VF,TOT}\cdot(1-\Gamma_{IN}))$
CSSF	$1+(1+\Gamma_{IN}-2\cdot A_{VF,TOT})/(A_{VF,TOT}^2(1-\Gamma_{IN}))+NEF\cdot(1+\Gamma_{IN})\cdot(1-A_{VF,TOT})^2/(A_{VF,TOT}^2(1-\Gamma_{IN}))$
A1	$1+NEF\cdot(1+\Gamma_{IN})/(1-\Gamma_{IN})-2/(A_{VF,TOT}\cdot(1-\Gamma_{IN}))$
CC CG	$1+(NEF/2)\cdot(1+\Gamma_{IN,D})/(1-\Gamma_{IN,D})+2/(A_{VF,TOT}\cdot(1-\Gamma_{IN,D}))$

Table 4.3: F as a function of the input reflection coefficient $\Gamma_{IN(,D)}$ and the gain $A_{VF,TOT,D}$.

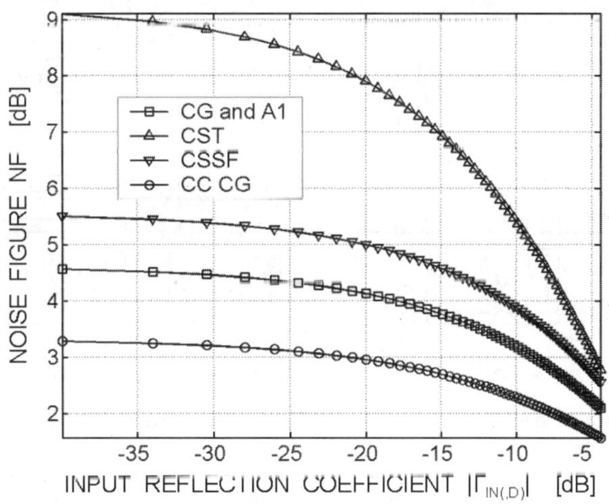

Figure 4.5: NF versus $|\Gamma_{IN,(D)}|$ for $|A_{VF,TOT}|=5$ (NEF=1.5).

4.5 Breaking the Trade-off via Negative Feedback

The trade-off between F and $Z_{IN}=R_S$ is broken when the noise factor can be made arbitrarily smaller than 1+NEF regardless the impedance matching requirement. Such operation can be performed exploiting properly negative feedback. In this respect, amplifiers exploiting non-energetic devices (e.g.: transformers) as feedback elements provide generally lower noise factors [7]. Furthermore, since a larger amount of feedback can be applied without taking signal power from the output, their linearity (and so the dynamic range) can be superior as well. In practice, the performance of these amplifiers critically depends on the availability of adequate transformers. Since wide-band transformers are difficult to integrate (particularly for low-cost digital CMOS using highly doped substrate [9]), transformer-feedback LNAs has been predominantly realised using high-quality wide-band discrete ferrite transformers for frequencies up to about 1GHz [5,6,8]. A sub-optimal solution is represented by lossless energetic feedback via capacitors and inductors [7]. However, their frequency dependent reactance makes more difficult the design of amplifiers with wide band response (e.g.: Z_{IN}=R_S). Ultimately, resistive or active feedback is perhaps the most practical solution. In this respect, complicated feedback arrangements with one or more loops are possible [7, 10]. However, single-loop circuits are simpler, easier to design and so more suitable for a high frequency design.

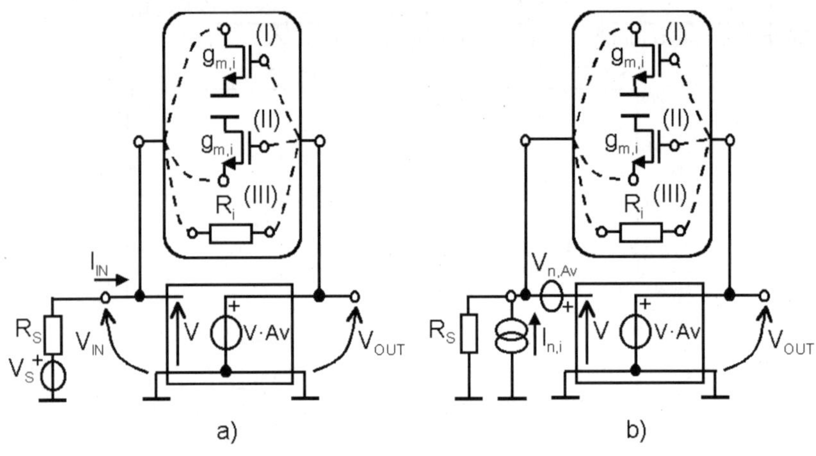

Figure 4.6: Wide-band negative feedback LNAs a) and its noise model b).

Figure 4.6a shows perhaps the simplest example of a single-loop wide band amplifier capable of achieving low F upon Z_{IN}=R_S. A generic V-I converter with transconductance $G_{m,i} \in \{1/R_i, g_{m,i}\}$ is exploited as a feedback network around a loop (voltage) amplifier with a gain $Av=V_{OUT}/V_{IN}$. The latter boost the voltage drop across the input of the V-I network, thereby allowing for an input impedance Z_{IN} that is significantly smaller than $1/G_{m,i}$ as shown in table 4.4. For this amplifier, the impedance matching requirement fixes the value of the product $G_{m,i}$ time the gain Av of the loop-amplifier.

LNA	$G_{m,i}$	Z_{IN}	Gain	EF_i	EF_i for $Z_{IN}=R_S$
Fig. 4.6a (I)	$g_{m,i}$	$1/(Av \cdot G_{m,i})$	Av>0		NEF/Av
Fig. 4.6a (II)	$g_{m,i}$	$1/((1-Av) \cdot G_{m,i})$	Av<0	$NEF \cdot G_{m,i} \cdot R_S$	NEF/(1-Av)
Fig. 4.6a (III)	$1/R_i$	$1/((1-Av) \cdot G_{m,i})$	Av<0		1/(1-Av)

Table 4.4: Feedback amplifiers transfer properties and noise factor F.

To see how this feedback arrangement can break the trade-off between F and $Z_{IN}=R_S$, consider the noise model of figure 4.6b. Here, the two main noise sources of the amplifier are shown: the noise current of the feedback device, $I_{n,i}$, in parallel to the input source[IX] and the (equivalent) input noise voltage of the loop amplifier, $V_{n,Av}$ (i.e. its equivalent noise current is assumed negligible). The trade-off with F is broken when the EF of both the feedback network and amplifier Av can be minimised upon $Z_{IN}=R_S$. Let's first consider the contribution of the feedback network. Its noise current $I_{n,i}$ adds to the source signal current I_S ($=V_S/R_S$, Norton model of the source). The output SNR, $SNR_{OUT,i}$, is equal to:

$$SNR_{OUT,i}=\frac{I_S^2}{I_{n,i}^2}=\frac{I_S^2}{4kT\cdot NEF\cdot G_{m,i}\cdot \Delta f}=\frac{SNR_{IN}}{NEF\cdot G_{m,i}R_S} \tag{4.3}$$

From equation (4.3), EF_i of the amplifier in figure 4.6 is obtained as shown in table 4.4. For $Z_{IN}=R_S$ we observe that:

- EF_i is smaller than 1 for any |Av|>NEF. This is because, via the feedback, a real input impedance $Z_{IN}=R_S$ requires a value of $G_{m,i}$ that is much smaller than $1/R_S$. Thus, $I^2_{n,i}$ can be much smaller than $I^2_{n,RS}$. For instance, for the case of a feedback resistor R_i, the noise current $I^2_{n,i}$ is (1-Av) times smaller than $I^2_{n,RS}$. Thus, the negative feedback acts in order *to transform* the equivalent input noise current *down* to a value lower than that provided by a resistor equal to R_S, while $Z_{IN}=R_S$ is kept.
- For the same Av, figure 4.6a (III) renders the lowest EF_i, provided NEF>1 holds.

The excess noise factor of the loop-amplifier, EF_{Av}, is calculated as follows:

- According to KCL and KVL laws, $V_{n,Av}$ is moved toward the input and output ports (see figure 4.7) paying attention to preserve the polarity of $V_{n,Av}$ once it is assigned.

[IX] In figure 4.6a (III), $I_{n,i}$ is connected between the input and output nodes. This is equivalent to a source $I_{n,i}$ connected from the input node to ground because the amplifier output node is driven by a voltage source with zero input impedance.

- The total output noise is calculated from the correlated noise contributions at the input and output ports.
- $SNR_{OUT,Av}$ is the ratio between the output signal and the total output noise.

According to the previous procedure, assuming an input referred noise voltage $V_{n,Av}$ for the loop amplifier, $SNR_{OUT,Av}$ can be written as follows:

$$SNR_{OUT,Av}=\begin{cases} \dfrac{V_S^2}{V_{n,Av}^2}=SNR_{IN}\cdot\dfrac{R_S}{R_{n,Av}} & ,(I) \\[2ex] \dfrac{V_S^2}{(1+g_{m,i}\cdot R_S)^2\cdot V_{n,Av}^2}=SNR_{IN}\cdot\dfrac{R_S}{R_{n,Av}}\cdot\dfrac{1}{(1+g_{m,i}\cdot R_S)^2} & ,(II) \\[2ex] \dfrac{V_S^2}{(1+\frac{R_S}{R_i})^2\cdot V_{n,Av}^2}=SNR_{IN}\cdot\dfrac{R_S}{R_{n,Av}}\cdot\dfrac{1}{(1+\frac{R_S}{R_i})^2} & ,(III) \end{cases} \qquad (4.4)$$

Where $R_{n,Av}=V_{n,Av}^2/(4kT\cdot\Delta f)$ is the equivalent noise resistance of the loop-amplifier Av.

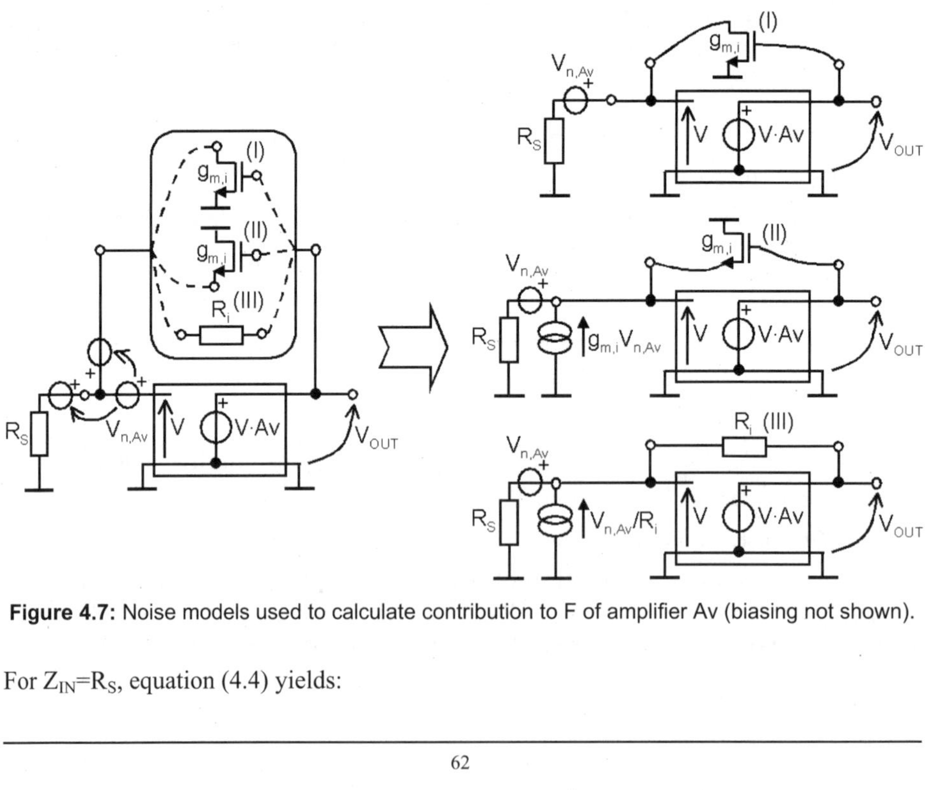

Figure 4.7: Noise models used to calculate contribution to F of amplifier Av (biasing not shown).

For $Z_{IN}=R_S$, equation (4.4) yields:

$$\mathrm{SNR}_{\mathrm{OUT,Av}}=\begin{cases}\mathrm{SNR}_{\mathrm{IN}}\cdot\dfrac{R_S}{R_{n,Av}} & ,\mathrm{(I)}\\ \mathrm{SNR}_{\mathrm{IN}}\cdot\dfrac{R_S}{R_{n,Av}}\cdot\dfrac{(1-Av)^2}{(2-Av)^2} & ,\mathrm{(II)\ and\ (III)}\end{cases} \tag{4.5}$$

$$\mathrm{EF}_{Av}=\begin{cases}\dfrac{R_{n,Av}}{R_S} & ,\mathrm{(I)}\\ \dfrac{R_{n,Av}}{R_S}\cdot\dfrac{(2-Av)^2}{(1-Av)^2} & ,\mathrm{(II)\ and\ (III)}\end{cases} \tag{4.6}$$

Equation (4.6) shows that:

- For a given Av, the value of EF_{Av} can be arbitrarily smaller than one as $R_{n,Av}$ drops below R_S. This is because the loop amplifier Av is not constrained by the matching requirement, thus the g_m of its input stage MOST can be chosen much larger than $1/R_S$. Thus, for $R_{n,Av}\rightarrow 0$ (i.e. $g_{m,Av}R_S\rightarrow\infty$), *the trade-off between F and $Z_{IN}=R_S$ is broken with a degree of de-coupling determined by the $G_{m,i}$ of the feedback network, which is in turn fixed by the gain Av and source impedance R_S.*
- For the same Av and $R_{n,AV}$, (I) renders the lowest contribution to F.

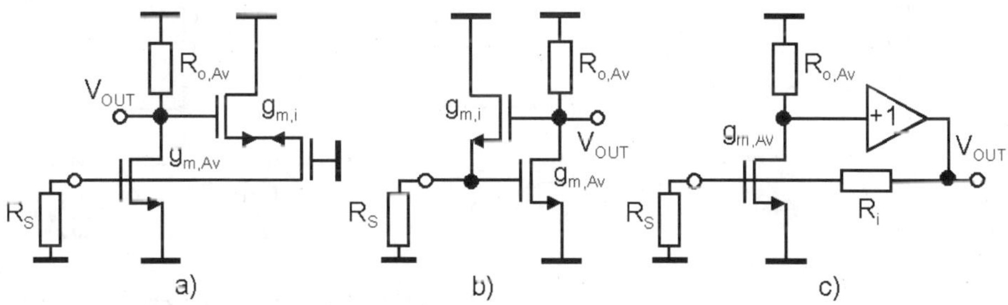

Figure 4.8: Implementations of the feedback LNAs in figure 4.6a (biasing not shown).

LNA	$R_{n,Av}/R_S$	F for $Z_{IN}=R_S$	F for $g_{m,Av}R_S\rightarrow\infty$
a		$1+NEF/A_V+R_{n,Av}/R_S$	$1+NEF/A_V$
b	$(NEF-1/A_V)/(g_{m,Av}\cdot R_S)$	$1+NEF/(1-A_V)+(R_{n,Av}/R_O)(2-A_V)^2/(1-A_V)^2$	$1+NEF/(1-A_V)$
c		$1+1/(1-A_V)+(R_{n,Av}/R_S)(2-A_V)^2/(1-A_V)^2$	$1+1/(1-A_V)$

Table 4.5: Expressions of F for the feedback LNAs shown in figure 4.8.

Figure 4.8 shows some elementary implementations of the generic amplifier of figure 4.6a (biasing not shown). A common-source amplifier (eventually) followed by a voltage buffer replaces the loop-amplifier Av. In case of figure 4.6a-(I), a differential pair ensures that the feedback is negative. Table 4.5 shows the expressions of their F. The latter are plotted in figure 4.9 versus the $g_{m,Av}R_S$ for Av=-10 and NEF=1.5.

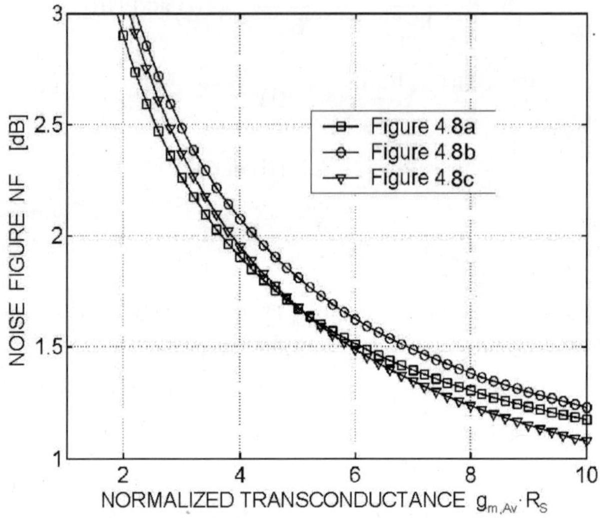

Figure 4.9: NF versus $g_{m,Av}R_S$ for $Z_{IN}=R_S$, Av=-10 and NEF=1.5.

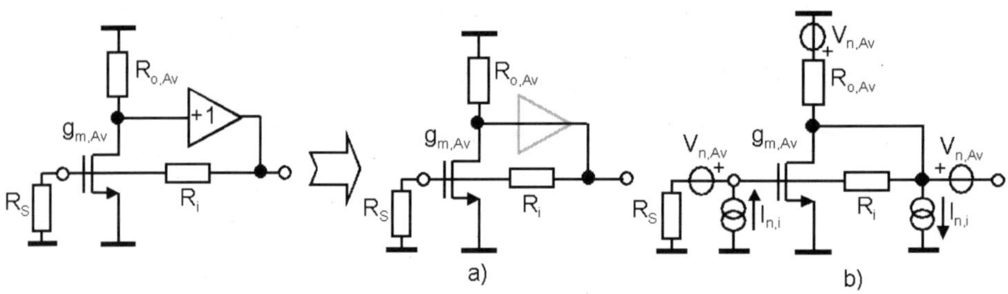

Figure 4.10: Alternative implementations of the feedback LNA in figure 4.8c (biasing not shown).

From this figure, one observes:

- NF drops to values well below 3dB for a properly large value of $g_{m,Av}R_S$. Ultimately, for $g_{m,Av}R_S \rightarrow \infty$, the lowest value of F is given by the *constant* contribution of the feedback device, which is determined by the gain A_v (see table 4.5).
- The amplifiers show essentially the same NF performance[X].

Figure 4.10a shows another implementation of the amplifier of figure 4.8c. Here, a short circuit replaces the voltage buffer. For this circuit, in order to provide voltage gain, resistor

[X] Actually, the NF of the LNA in figure 4.8c is somewhat higher because the buffer adds extra noise and the finite output impedance, which decreases the value of R_i.

$R_{o,Av}$ must then conduct most of the signal current delivered by the input MOST. Another way to look at this circuit is to regard it as a CSSF amplifier of figure 4.2c plus a resistive output termination equal to $R_{o,Av}$. In this view, the output termination in combination with the shunt-feedback resistor forms a current divider such that only a fraction of the drain current is fed back to the input. Consequently, F can be lower than 2 because $g_{m,Av} \cdot R_S$ is larger than 1 in order to provide $Z_{IN}=R_S$.

The feedback LNAs analysed in this paragraph suffer from important drawbacks:

- Sufficient gain and GHz bandwidth often mandates the use of multiple-cascaded stages *within* the feedback loop (e.g.: 2 stages for the amplifier in figure 4.8c). A wide-band amplifier with a loop transfer function characterised by multiple-poles (e.g.: 3 poles for the amplifier in figure 4.8c) is prone to instability.
- Furthermore, good linearity is subordinated to the availability of a sufficiently large loop gain. The latter is typically scarce at RF frequency OR it may lead to conflicting requirements. In general, the linearity of the feedback LNA in figure 4.6a isn't significantly better than that of its loop-amplifier Av. This is because, for $Z_{IN}=R_S$, the loop-gain $A_{LOOP}=-Av \cdot G_{m,i}R_S/(G_{m,i}R_S+1)$ ($A_{LOOP}=-Av \cdot G_{m,i}R_S$ for (I)) is always lower or equal to one regardless the gain Av. Since amplifier Av can consist of several cascaded stages with most of the gain in the first one (i.e. for best noise performance), linearity can be poor [11]. For the amplifier of figure 4.6a, low NF can occur at the price of an unsatisfactory linearity.
- The value of Z_{IN} depends on all the circuit paramcters: $g_{m,Av}$, $R_{o,Av}$ and $G_{m,i}$. Therefore:
  - Z_{IN} is rather sensitive to device parameter variations (e.g.: proccss-spread).
  - Variable gain at constant impedance match is not straightforward because Z_{IN} and A_{VF} are directly coupled.
- The reverse isolation of the amplifier in figure 4.10b is often insufficient (table 4.6).

LNA	$A_{VR}=V_{IN}/V_{OUT}$	A_{VR} for $Z_{IN}=R_S$
Fig. 4.8a	$-g_{m,i}R_S$	1/Av
Fig. 4.8b	$g_{m,i}R_S/(g_{m,i}R_S+1)$	1/(2-Av)
Fig. 4.10a	$R_S/(R_S+R_i)$	

Table 4.6: Reverse voltage gain.

4.6 The Noise Cancelling Technique

In this paragraph, a novel wide-band low-noise technique is presented, which is able to de-couple F from $Z_{IN}=R_S$ without needing (intended) feedback or degrading the quality of the matching. The underlying idea of this technique is that impedance matching and F are decoupled by *cancelling* properly the output noise from the *matching device*. After all, if the matching device does not contribute output noise, it does not affect F too. To understand how this can be done recall the noise-cancelling mechanism for amplifier A2 in chapter 3. Figure 4.11a shows the path of the noise current of the matching device, $I_{n,i}$.

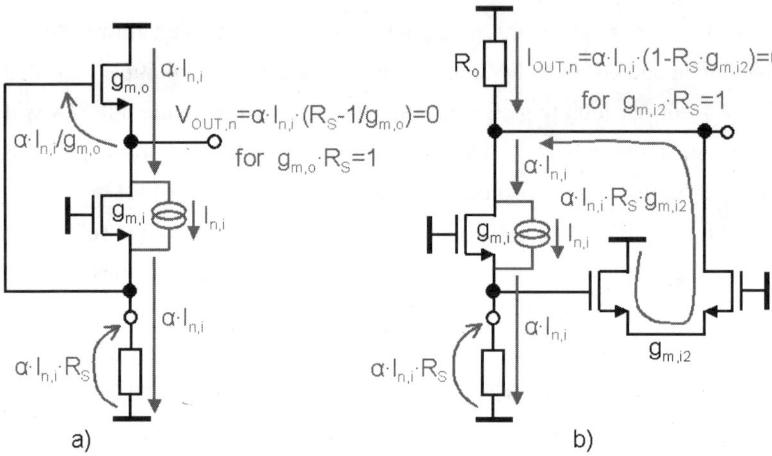

Figure 4.11: Wide-band noise-cancelling LNAs (biasing not shown): cancelling of the matching device output noise is indicated.

A noise current $\alpha(g_{m,i},R_S)\cdot I_{n,i}$ with $0<\alpha(g_{m,i},R_S)=1/(1+g_{m,i}\cdot R_S)<1$ flows out of the matching device through the source resistance R_S and the upper MOST. For $g_{m1}R_S=1$ (i.e. matching), this current is equal to $\frac{1}{2}\cdot I_{n,i}$ (i.e. the other half flows into the device itself). The noise current $\alpha(g_{m,i},R_S)\cdot I_{n,i}$ gives rise to two *fully correlated* noise voltages $R_S\cdot\alpha(g_{m,i},R_S)\cdot I_{n,i}$ and $\alpha(g_{m,i},R_S)\cdot I_{n,i}/g_{m,o}$ across R_S and the gate-source terminals of the upper MOST respectively. As the output noise, $V_{OUT,n,i}$, is the *instantaneous* difference of the previous contributions as $V_{OUT,n,i}=\alpha(g_{m,i},R_S)\cdot I_{n,i}\cdot(R_S-1/g_{m,o})$, exact output *noise voltage cancellation* occurs for $g_{m,o}\cdot R_S=1$. In a similar manner, noise cancelling can also be performed in the current domain as shown in figure 4.11b. Here, the noise current outing from the matching device, $\alpha(g_{m,i},R_S)\cdot I_{n,i}$, mirrors a noise current $R_S\cdot g_{m,i2}\cdot\alpha(g_{m,i},R_S)\cdot I_{n,i}$ into the differential pair, which subtracts from the former at the output. The output noise current, $I_{OUT,n,i}=\alpha(g_{m,i},R_S)\cdot I_{n,i}\cdot(1-R_S\cdot g_{m,i2})$, is then zero for $g_{m,i2}\cdot R_S=1$. In this case, the matching device does not contribute output noise because its noise current flows in a loop that does not include the load R_o.

In both the amplifiers in figure 4.11, two feed-forward transfer paths from the input node to the output can be distinguished. One path is through the matching device, the other via a voltage-sensing amplifier (i.e. the CD stage in figure 4.11a and the differential pair in figure 4.11b). Since the matching device does not contribute output noise and signal components along the two paths *do add* in phase at the output (see figure 4.12), the trade-off between F and $Z_{IN}=R_S$ seems to be broken.

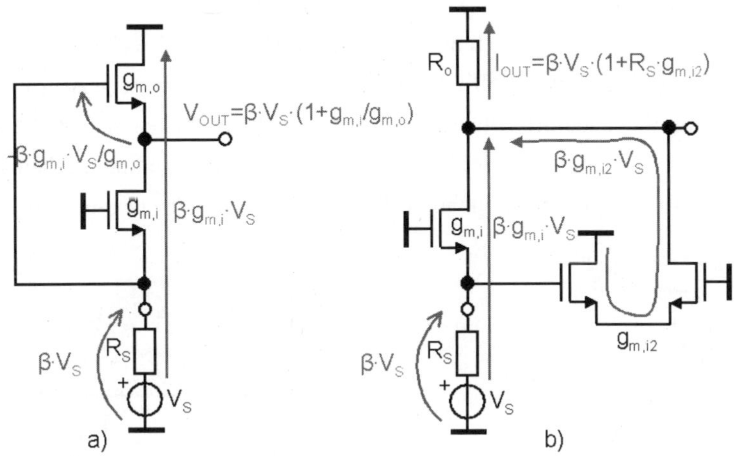

Figure 4.12: Addition of signals for the LNAs in figure 4.11.

Unfortunately, we are not done yet because for both the amplifiers in figure 4.11, the noise cancellation conditions $g_{m,o}\cdot R_S=1$ and $g_{m,i2}\cdot R_S=1$ constraint the g_m of the upper MOSFET (figure 4.11a) and the differential pair (figure 4.11b) an so their contribution to the noise factor. This is evident looking at their output SNR, $SNR_{OUT,o(i2)}$:

$$SNR_{OUT,o(i2)}=\begin{cases}\dfrac{\dfrac{V_S^2\cdot\beta^2\cdot(1+\frac{g_{m,i}}{g_{m,o}})^2}{4kT\cdot NEF\cdot\Delta f}}{g_{m,o}} & ,\text{fig. 4.11a}\\[2ex] \dfrac{V_S^2\cdot\beta^2\cdot(g_{m,i}+g_{m,i2})^2}{4kT\cdot NEF\cdot g_{m,i2}\cdot\Delta f} & ,\text{fig. 4.11b}\end{cases} \tag{4.7}$$

and $0<\beta=\beta(g_{m,i}, R_S)<1$ is the transfer function from V_S to the LNA input. For $Z_{IN}=1/g_{m,i}=R_S$ (i.e. $\alpha=\beta=1/2$), equation (4.7) is written in term of SNR_{IN} as:

$$SNR_{OUT,o(i2)} = \begin{cases} \dfrac{SNR_{IN}}{NEF} \cdot \dfrac{(g_{m,o} \cdot R_S + 1)^2}{4 \cdot g_{m,o} \cdot R_S} & \text{, fig. 4.11a} \\ \dfrac{SNR_{IN}}{NEF} \cdot \dfrac{(1 + g_{m,i2})^2}{4 \cdot g_{m,i2} \cdot R_S} & \text{, fig. 4.11b} \end{cases} \tag{4.8}$$

For $g_{m,o} \cdot R_S = 1$ and $g_{m,i2} \cdot R_S = 1$, equation (4.8) yields to $EF_{o(i2)} = NEF$. Thus, noise cancelling constraints F to be larger than 1+NEF because of the noise of the voltage sensing devices. In other words, the trade-off between EF_i and $Z_{IN} = R_S$ is broken but at the price of introducing the same trade-off with $EF_{o(i2)}$. Clearly, the noise cancelling mechanism needs to be improved to overcome this limitation. This will be done in the next paragraph.

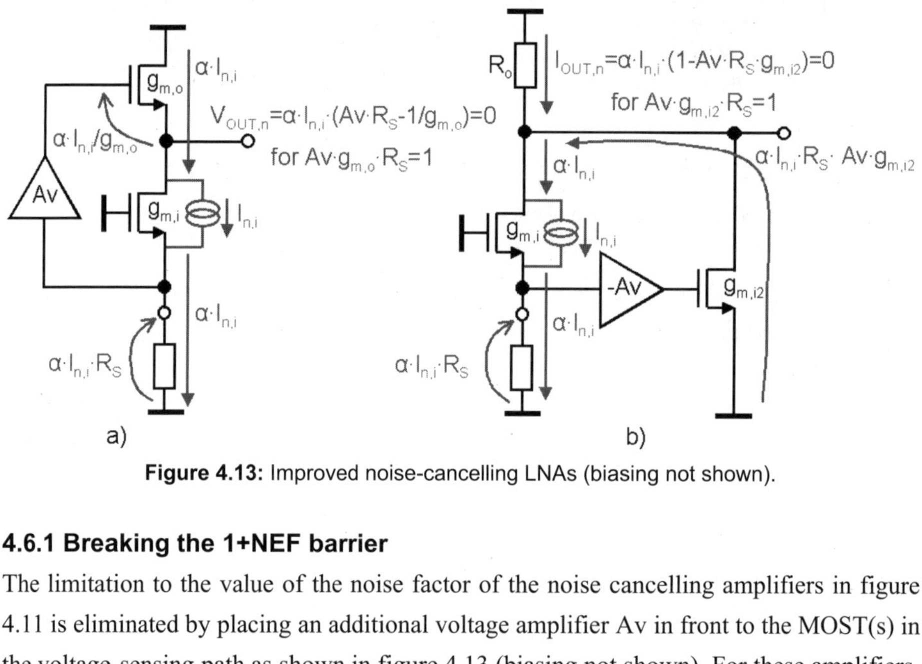

Figure 4.13: Improved noise-cancelling LNAs (biasing not shown).

4.6.1 Breaking the 1+NEF barrier

The limitation to the value of the noise factor of the noise cancelling amplifiers in figure 4.11 is eliminated by placing an additional voltage amplifier Av in front to the MOST(s) in the voltage-sensing path as shown in figure 4.13 (biasing not shown). For these amplifiers, the noise cancelling equations can then be written as:

$$\begin{aligned} V_{OUT,n,i} &= \alpha(g_{mi}, R_S) \cdot I_{n,i}\left(A_V R_S - 1/g_{m,o}\right) = 0 \\ I_{OUT,n,i} &= \alpha(g_{mi}, R_S) \cdot I_{n,i}\left(1 - A_V R_S g_{m,i2}\right) = 0 \end{aligned} \tag{4.9}$$

In this case, noise cancelling occurs for $g_{m,o}\cdot Av\cdot R_S=1$ and $g_{m,i2}\cdot Av\cdot R_S=1$. Compared to the amplifiers in figure 4.11, the auxiliary amplifier Av brings in the design the extra degree of freedom (i.e. its gain), which can be exploited to decrease F well below 1+NEF barrier without compromising the matching $Z_{IN}=R_S$. This is possible because:

- The EF of the device following amplifier "Av" is negligible for Av>>1.
- EF_{Av}, (i.e. the main contribution to F upon noise cancellation), can be made arbitrarily smaller than one because this amplifier is *not* constrained by impedance matching.

From equation (4.9), two important characteristics of noise cancelling are:

- The noise cancelling condition is independent on the $g_{m,i}$ of the matching device and so on the quality of the impedance matching. This is because any change in the value of $g_{m,i}$ result in identical variations of the two feed-forward noise transfers. Indeed, the value of $g_{m,i}$ determines only the amount of output noise to be cancelled.
- Noise cancelling depends on the absolute value of the *real* impedance of the source, R_S (e.g.: the input impedance seen "looking into" a properly terminated coax cable). Moreover, in general, R_S does not affect equally the two feed-forward noise transfers.

Compared to the feedback LNAs of section 4.5, noise cancelling offers some advantages:

- It can be *free* of (intended) feedback, thus instability risks are relaxed.
- Z_{IN} depends only on the g_m of one matching MOST. The consequences are twofold:
  - Z_{IN} is less sensitive to device parameter variations (e.g.: due to process spread).
  - Variable gain at constant $Z_{IN}=R_S$ is rather straightforward (e.g.: in figure 4.13b, by varying the value resistor R_o).

Furthermore, it is worth noting that any undesired small signal that can be modelled as a current source between the drain and source of the matching device is cancelled too. This includes, for instance, 1/f noise and thermal noise of the distributed resistance of the gate

4.6.2 Noise cancelling generalisation

In the previous paragraph, the noise cancelling concept has been described with regard to the specific amplifier circuits in figures 4.11 and 4.13. In this paragraph, it will be shown that noise cancelling is indeed a general technique that can have more circuit implementations. Figure 4.14a shows the conceptual block-diagram of a wide-band amplifier exploiting noise cancelling.

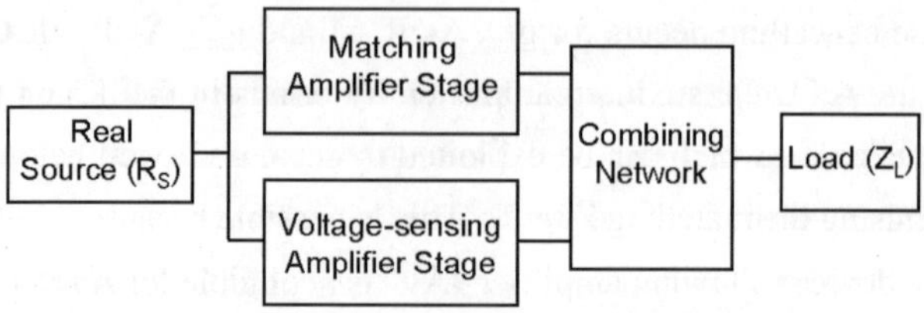

a)

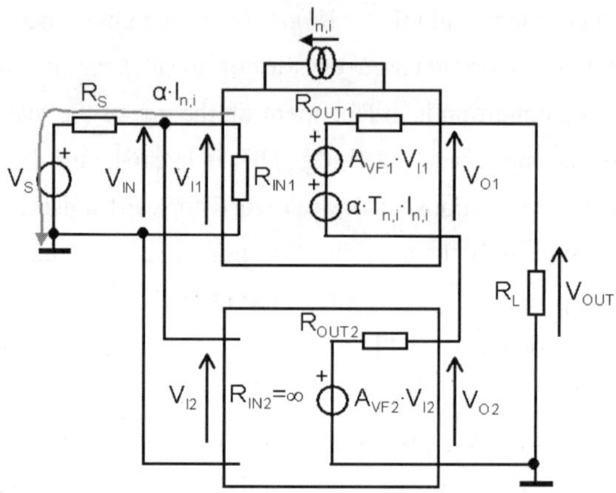

b)

Figure 4.14: Representation of a generic noise cancelling wide-band LNA: conceptual block diagram and (b) Two-port model.

It consists of the following functional blocks:

- An amplifier stage *to match* the real source impedance, R_S.
- An amplifier stage *to sense* the voltage drops (both signal and noise) across R_S.
- A network combining the outputs of the two amplifiers.

The generic description in figure 4.14a can be translated into a more useful two-port model of figure 4.14b. The latter uses two two-port amplifiers connected to the output in feed-forward. Two feed-forward paths are needed to cancel the output noise from the matching device while adding in-phase signal components. In this model, the series connection of the output ports represents the combining network. One two-port provides the LNA input impedance, $R_{IN}=R_{IN1}$ (i.e. matching two-port). The other one allows noise cancelling to take place at the output by sensing (i.e. $R_{IN2}=\infty$) the matching device noise voltage across R_S without loading it (i.e. voltage sensing two-port). For noise cancelling to occur, the following conditions must hold[XI]:

[XI] For simplicity $R_L=\infty$ is assumed. However, the conclusions of this paragraph hold for any value of R_L.

- **Signals add.** The voltage gain, $A_{VF,TOT}=V_{OUT}/V_S=\beta(R_S,g_{m,i})\cdot(V_{OUT}/V_{IN})$, can be written as: $\beta(R_S,g_{m,i})\cdot(A_{VF1}+A_{VF2})$, with $A_{VF1(2)}=V_{OUT}/V_{IN}$ for $A_{VF2(1)}=0$[XII]. Since signals must add in-phase to the output, $\text{sign}(A_{VF1})=\text{sign}(A_{VF2})$ must holds otherwise the combining network must perform the required sign inversion.
- **Noise Cancels.** The noise current, $I_{n,i}$, of the matching device produces output noise via two paths. The first contribution through the matching two-port has a transfer function $H_{n,i,1}=\alpha(R_S,g_{m,i})T_{n,i}$ where $T_{n,i}$ is the transfer from $I_{n,i}$ to the output through the matching two-ports when the input of the voltage sensing two-ports is grounded (i.e. $T_{n,i}=V_{OUT}/I_{n,i}$ for $A_{VF2}=0$). The second contribution via the source resistance R_S and the voltage sensing two-port has a transfer function $H_{n,i,2}=\alpha(R_S,g_{m,i})R_S A_{VF2}$. The total noise transfer, $H_{n,i}$, from $I_{n,i}$ to the output is then:

$$H_{n,i}=H_{n,i,1}+H_{n,i,2}=\alpha(g_{mi},R_S)\cdot(R_S\cdot A_{VF2}+T_{n,i})$$

Noise cancelling at the output, $H_{n,i}=H_{n,i,1}+H_{n,i,2}=0$, leads to the following condition:

$$H_{n,i,1}=-H_{n,i,2} \quad \Leftrightarrow \quad T_{n,i}=-R_S\cdot A_{VF2} \quad \forall\alpha$$

The noise transfer $T_{n,i}$ and gains A_{VF2} (and A_{VF1}) must have *different* signs. In general, both the feed-forward transfers $H_{n,i,1}$ and $H_{n,i,2}$ may depend on R_S and this dependence does not need to be the same.

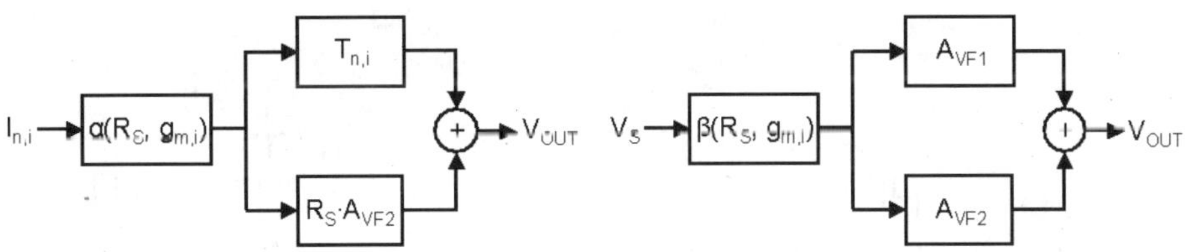

Figure 4.15: Block diagram of the matching device noise and input signal transfer.

The feed-forward nature of the noise cancelling technique is further highlighted in figure 4.15, where the transfers of the signal and the matching-device noise are represented. According to the given modelling, alternative implementations of noise-cancelling amplifiers can be constructed. For instance, some of the elementary amplifiers in figure 4.2 can be used as matching two-port stage. The implementation of the voltage sensing

[XII] The input of the two-port is connected to ground when evaluating the gain of the other two-port.

two-port (e.g.: a CS or a differential pair gain stage) must be chosen according to the sign of the gain of the matching stage. Figure 4.16 shows a sample of noise-cancelling amplifiers. Their model parameters and the noise-cancelling equation are shown in table 4.8. Note as only for amplifiers f and h, $T_{n,i}$ depends on R_S, anyhow the dependence is not the same as for the other feed-forward noise transfer.

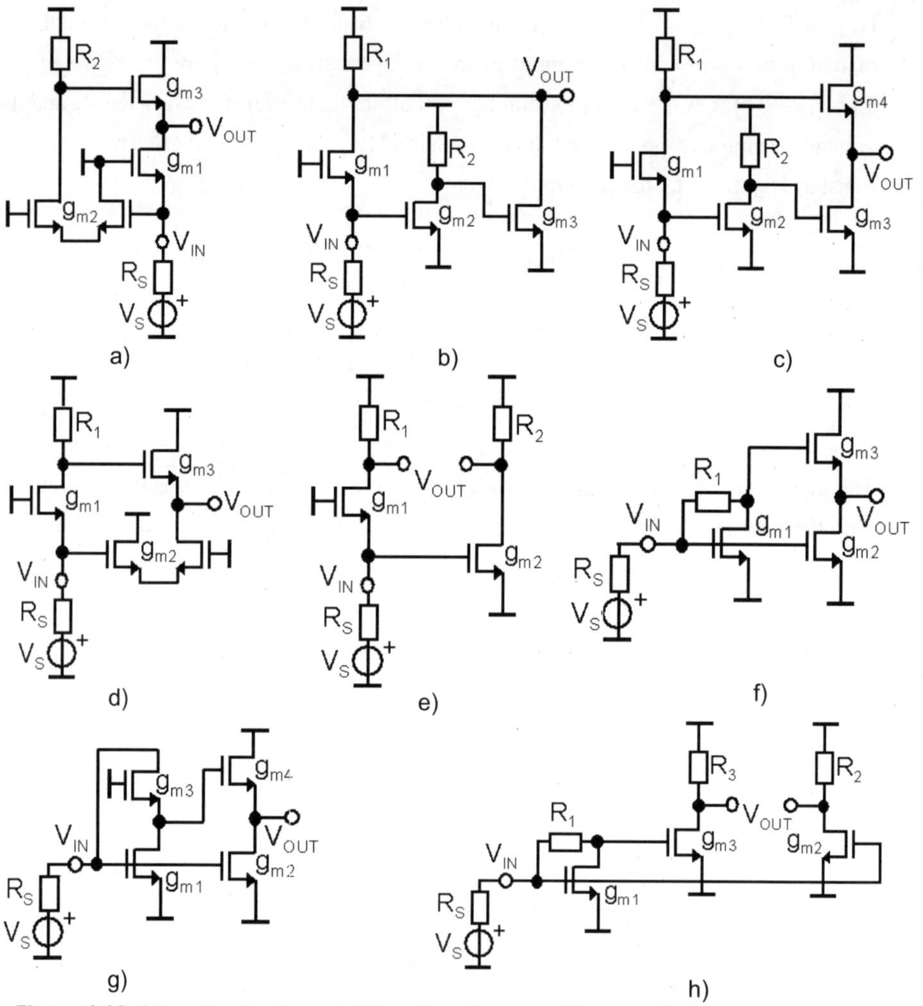

Figure 4.16: Alternative implementations of a noise-cancelling amplifier (biasing not shown).

LNA	Z_{IN1}	Z_{OUT}	A_{VF1}	A_{VF2}	$T_{n,i}$	$H_{n,i}=T_{n,i}+R_S \cdot A_{VF2}$
a	$1/g_{m1}$	$1/g_{m3}$	g_{m1}/g_{m3}	$g_{m2}R_2$	$-1/g_{m3}$	$R_S g_{m2}R_2 - 1/g_{m3}$
b	$1/g_{m1}$	R_1	$g_{m1}R_1$	$g_{m2}R_2 g_{m3}R_1$	$-R_1$	$(R_S g_{m2}R_2 g_{m3} - 1) \cdot R_1$
c	$1/g_{m1}$	$1/g_{m4}$	$g_{m1}R_1$	$g_{m2}R_2 g_{m3}/g_{m4}$	$-R_1$	$R_S g_{m2}R_2 g_{m3}/g_{m4} - R_1$
d	$1/g_{m1}$	$1/g_{m3}$	$g_{m1}R_1$	g_{m2}/g_{m3}	$-R_1$	$R_S g_{m2}/g_{m3} - R_1$
e	$1/g_{m1}$	R_1+R_2	$g_{m1}R_1$	$g_{m2}R_2$	$-R_1$	$R_S g_{m2}R_2 - R_1$
f	$1/g_{m1}$	$1/g_{m3}$	$1-g_{m1}R_1$	$-g_{m2}/g_{m3}$	R_S+R_1	$R_S g_{m2}/g_{m3} - (R_S+R_1)$
g	$1/g_{m1}$	$1/g_{m4}$	$-g_{m1}/g_{m3}$	$-g_{m2}/g_{m4}$	$1/g_{m3}$	$1/g_{m3} - R_S g_{m2}/g_{m4}$
h	$1/g_{m1}$	R_3+R_2	$(1-g_{m1}R_1)g_{m3}R_3$	$g_{m2}R_2$	$-(R_S+R_1)g_{m3}R_3$	$R_S g_{m2}R_2 - (R_S+R_1)g_{m3}R_3$

Table 4.8: Two-port model parameters of the noise cancelling amplifiers in figure 4.16

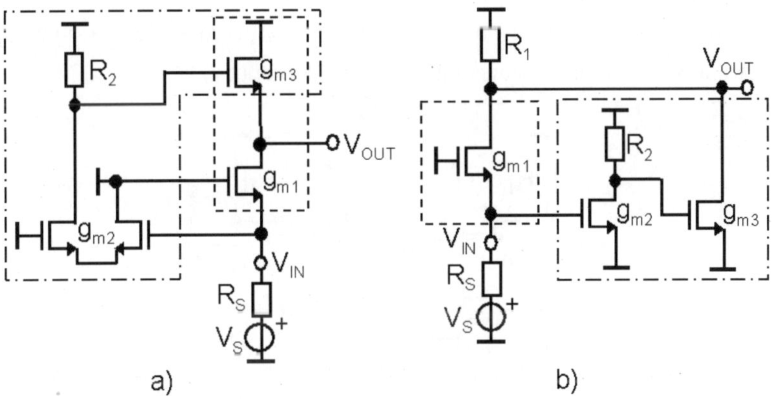

Figure 4.17: Two-port model of the amplifiers "a" and "b" in figure 4.16 (biasing not shown).

The two-ports in figure 4.14b do not necessarily correspond to physically separate circuits. For instance, amplifier of figure 4.16a can be separated as in figure 4.17a. The dashed-line box surrounding the CG stage defines the matching two-port with $Z_{IN}=Z_{IN1}=1/g_{m1}$. The dot-line-dot box is the second two-port. Note the double-role played by the MOST g_{m3}. It acts as a load of the CG gain stage and it buffers the signal voltage provided by the differential-pair gain stage to the output as well. Thus, both the two-ports in figure 4.14b share MOST g_{m3}. On the other hand, the amplifier of figure 4.16b can be split into two separate two-port circuits (figure 4.17b) because noise cancelling is achieved through the addition of currents at the output node. A different situation occurs for the amplifiers of figure 4.16e and figure 4.16h. In this case, the two two-ports can be physically separated because the addition of voltages at the output arises from taking a differential output. In addition, for amplifiers 'e' and 'h', noise-cancellation depends on the *ratio* of their load R_1/R_2 and R_3/R_2 respectively. These resistors can be replaced by a pair of noiseless passive devices such as capacitors or even inductors without affecting the noise-cancellation. In this case, for the same power, the noise figure can be lower at the price of a frequency-dependent gain response.

Not all the circuits in figure 4.16 can be considered new. The 2-MOSTs core of amplifier 4.16e is a known transconductor circuit [12], which was recently used a transconductor for a double-balanced mixer [13]. In both cases, apparently, the existence of noise cancelling was not recognised and therefore not exploited into the design.

Balanced noise-cancelling LNAs can be obtained from figure 4.16 by exploiting:

- Two equal and separate single-ended (SE) input single-ended output noise-cancelling amplifiers. In this case, noise cancelling occurs at each of the outputs.
- Two equal single-ended noise-cancelling amplifiers with cross-coupled signal-paths. In this case, noise cancelling occurs at the differential output.

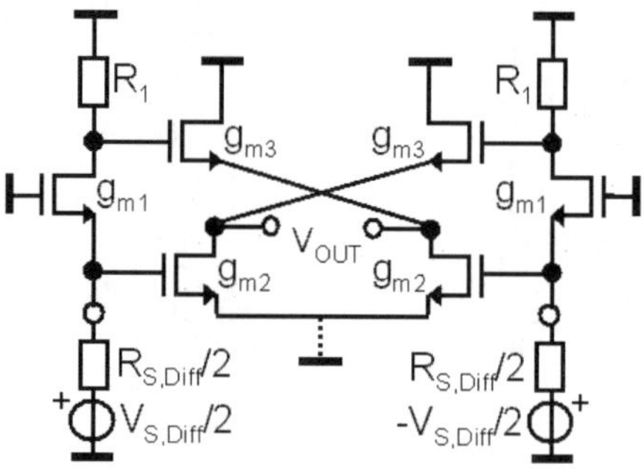

Figure 4.18: Balanced implementation of the amplifier in figure 4.16d exploiting cross coupling.

Figure 4.18 shows an example of a balanced noise-cancelling LNA based on the amplifier in figure 4.16d, which exploits cross coupling in the voltage sensing two-port. It uses a more power efficient CS stage instead of a differential pair (i.e. the g_m/I_D of a CS stage is 4 times larger than that of the differential pair stage in figure 4.16d). This leads to a lower F for the same power consumption and gain or less power for the same F. Capacitive input cross coupling can also be used to halve the power dissipated in the matching stage. This power can be spend to decrease F further.

4.6.3 Intuitive analysis noise cancelling

In this paragraph, an intuitive approach to noise cancellation is described. We have seen that in order for noise cancelling to occur, the transfer function of the matching device noise $T_{n,i}$ and the gains A_{VF2} and A_{VF1} must have *different* signs. This suggests that an intuitive manner to analyse how noise cancelling occurs is by looking at the *correlation* between the sign of the *instantaneous* signal and noise voltage at the input and output nodes of the matching two-port. For instance, consider amplifiers in figure 4.13a and 4.16f. Figure 4.19a and 4.19b sketches the signal and the matching device noise voltage at the input node X and output node Y of amplifier in figure 4.12a (i.e. Av=0). Since the gain of the CG amplifier is positive, voltage at nodes X and Y have different amplitude and *equal* sign (figure 4.19a). On the other hand, the noise current of the matching device, $\alpha(R_S,g_{m,i})\cdot I_{n,i}$, generates two instantaneous noise voltages at nodes X and Y that have different amplitude but *opposite* sign (figure 4.19b). This different correlation between the sign of the signal and noise voltage at nodes X and Y enables the possibility to cancel the noise of the matching device while simultaneously adding the signal components.

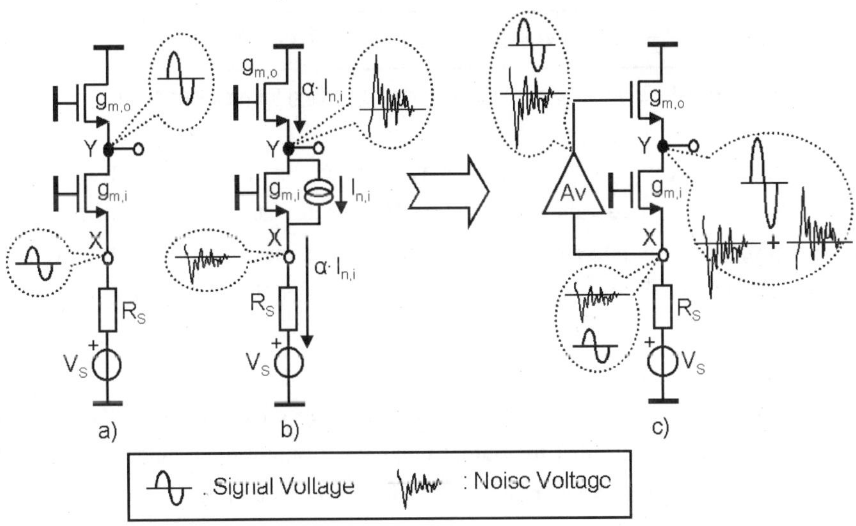

Figure 4.19: Intuitive analysis of noise cancelling for amplifier 4.12a.

This is achieved delivering to a new output the sum of the voltage at node Y plus the voltage at node X multiplied by a *positive* scaling factor. Clearly, a proper value of this scaling factor leads to the output noise voltage-cancellation. In figure 4.19c, this operation is performed by an auxiliary voltage-sensing amplifier "Av" connected between node X and the gate of the load MOST, which creates a second feed-forward path to the output. By circuit inspection, it is found that noise cancelling occurs for $V_{OUT,n,1}=V_{X,n,1}\cdot Av+V_{Y,n,1}=\alpha(R_S,g_{m,i})\cdot I_{n,i}\cdot(R_SA_V-1/g_{m,o})=0$ and so $A_VR_S=1/g_{m,o}$. For instance, amplifier "Av" can be implemented as shown in figure 4.16a.

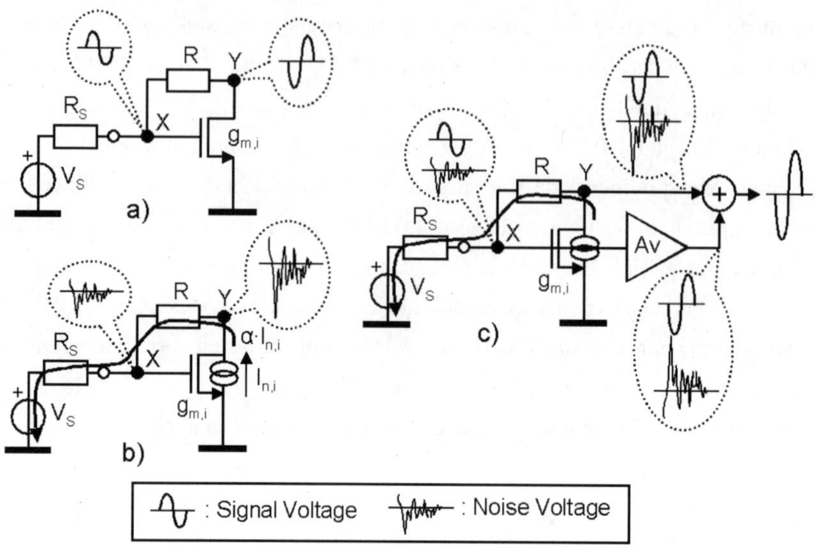

Figure 4.20: Intuitive analysis of noise cancelling for amplifier 4.16f.

The same reasoning can now be used to analyse the amplifier in figure 4.16f. Shown in figure 4.20a and figure 4.20b, the signal voltage at nodes X and Y have *opposite* sign (i.e. negative gain), while the noise voltage at the same nodes have *equal* sign (i.e. R_S and R are in series). Output noise cancelling is now performed by delivering the weighted difference, $V_{OUT,n,1}=V_{X,n,1}\cdot Av+V_{Y,n,1}$, of the voltage at node X and Y to the output as shown in figure 4.20c. Here, the cancellation condition is $Av=-V_{Y,n,1}/V_{X,n,1}=-1-R/R_S$. For $Z_{IN}=R_S$, the minimum *negative* gain Av rendering the cancellation is smaller than –2 (i.e. $1-R/R_S<0 \rightarrow R/R_S>1$). Note that the output noise contribution of R cannot be cancelled. This is easily seen by splitting its noise current $I_{n,R}$ into two correlated noise source from node X to ground and node Y to ground. The latter is cancelled together with the noise current of the matching device. The former cannot be distinguished from the input signal, so $EF_R=I^2_{n,RS}/I^2_{n,R}=R_S/R$. Amplifier "Av" is easiest implemented as shown in figure 4.16f.

The noise-cancelling amplifier in figure 4.20c has been designed in a standard 0.25μm CMOS process. The matching-stage provides $R_{IN}\approx 50\Omega$ for a $V_{GS}-V_{T0}\approx 250mV$.

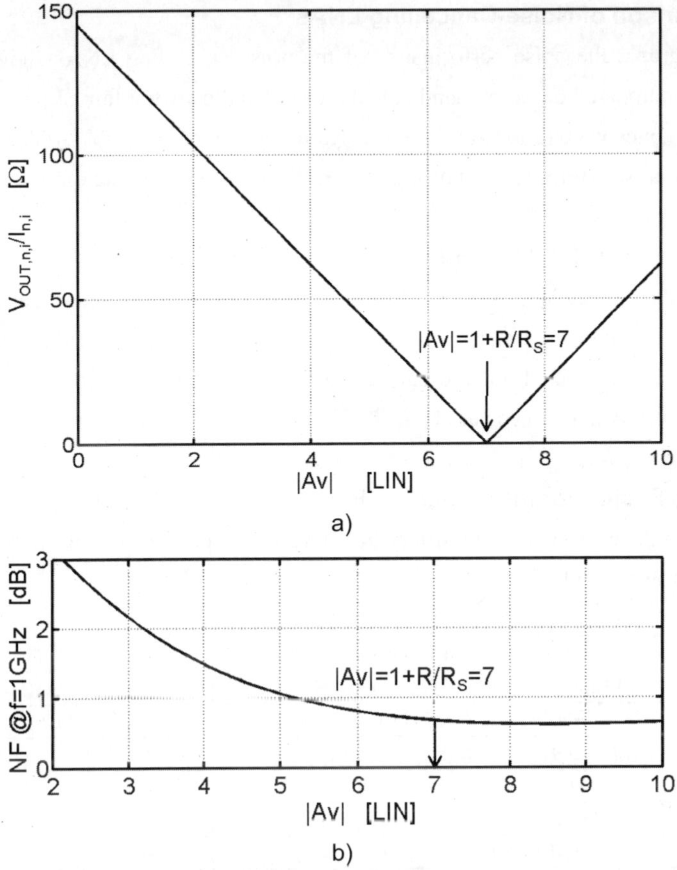

Figure 4.21: (a) Simulated transfer function from the noise current $I_{n,i}$ to the output versus the gain |Av| @1GHz and (b) Simulated noise figure versus the gain |Av| @1GHz.

A shunt-feedback resistor R equal to 300Ω renders a voltage gain ($=V_Y/V_X$) of 12.8dB for the matching stage. The voltage-sensing amplifier is implemented as an ideal voltage controlled voltage source of gain Av. Figure 4.21a shows the simulated transfer function from the matching-device noise current, $I_{n,i}$, to the LNA output $V_{OUT,n,i}$ versus |Av| at 1GHz (i.e. about 1/15 of the –3dB bandwidth). Clearly, in agreement with the hand calculations, exact noise cancellation occurs for $|Av|=1+R/R_S= 1+300/50=7$. The LNA noise figure is plotted in figure 4.21b versus |Av| at 1GHz. The noise figure drops from a maximum of 6dB at Av=0, (i.e. NF of the matching-stage standalone) as |Av| increases. NF is about 0.65dB for |Av|=7 that is the contribution of R (i.e. 0.67dB using $F=1+R_S/R$).

4.7 Comparison of Noise-Cancelling LNAs

In this paragraph, the noise performance of the noise cancelling LNAs shown in figure 4.16 will be compared based on hand calculations. To allow for a fair comparison of their noise performance we will look at F for $Z_{IN}=R_S$ and given gain $A_{VF}=V_{OUT}/V_{IN}$ versus:

- The (normalised) transconductance $g_{m2}\cdot R_S$ of the device mainly determining F.
- Power consumption.
- Power consumption & bias noise.

Source impedance matching is generally required in order to avoid signal reflections on a cable or alterations of the transfer characteristics of the RF filter preceding the LNA [1]. On the other hand, a fixed voltage gain ensures equal contribution to F from the stages following the LNA in the receiver chain[XIII].

4.7.1 Noise Factor versus device $g_{m2}R_S$

Using table 4.8, the F of the amplifiers in figure 4.16 upon noise cancellation has been calculated as shown in table 4.9.

LNA	F for $Z_{IN}=R_S$ and fixed A_{VF}	Condition	F for $g_mR_S\to\infty$
a	$1+2\cdot NEF/A_{VF}+(NEF+2/A_{VF})/(g_{m2}R_S)$	-	$1+2\cdot NEF/A_{VF}$
b	$1+2\cdot(1+NEF)/A_{VF}+(NEF+2/A_{VF})/(g_{m2}R_S)$	$\lvert A_2\rvert=A_{VF}/2$	$1+2\cdot(1+NEF)/A_{VF}$
c	$1+2/A_{VF}+(NEF+2/A_{VF})/(g_{m2}R_S)+8\cdot NEF/(g_{m3}R_SA_{VF}^2)$	$\lvert A_2\rvert=A_{VF}/2$	$1+2/A_{VF}$
d	$1+2/A_{VF}+NEF\cdot(1+2/A_{VF})/(g_{m2}R_S)$	-	$1+2/A_{VF}$
e	$1+2/A_{VF}+(NEF+2/A_{VF})/(g_{m2}R_S)$	-	$1+2/A_{VF}$
f	$1-2/A_{VF}+NEF\cdot(8-6\cdot A_{VF}+A_{VF}^2)/(g_{m2}R_SA_{VF}^2)$	-	$1-2/A_{VF}$
g	$1-2\cdot NEF/A_{VF}+NEF\cdot(1-2/A_{VF})/(g_{m2}R_S)$	-	$1-2\cdot NEF/A_{VF}$
h	$1-2/A_{VF}+4\cdot(NEF\cdot(1-A_{VF}/2)^2+1-A_{VF}/2)/(g_{m2}R_SA_{VF}^2)$ $+4\cdot(1+NEF)/(g_{m3}R_SA_{VF}^2)$	$\lvert A_3\rvert=1$	$1-2/A_{VF}$

Table 4.9: Noise factor F of the LNAs in figure 4.16 ($\lvert A_2\rvert=g_{m2}R_2$, $\lvert A_3\rvert=g_{m3}R_3$).

From table 4.9, we observe that:

- For fixed A_{VF}, F decreases as the normalised device $g_{m2}R_S$ increases above one.
- For $g_{m2(3)}\cdot R_S\to\infty$, the minimum value of F (i.e. F_{min}) is limited by the contribution of the matching stage load, which is determined by A_{VF} (table 4.9, most right column).
- The lowest value of F_{min} occurs for those LNAs whose matching stage is resistively loaded (i.e. R provides NEF times less noise power than a sub-micron MOST having $g_m=1/R$). An exception is represented by amplifier 4.16b. In this case, the noise

[XIII] Strictly speaking, a constant *available* gain, G_{av}, should be imposed. However, when the (second-stage) equivalent input noise current 'i_n' is negligible, G_{av} can be replaced by the voltage gain.

cancelling condition for $|A_2|=A_{VF}/2$ constraints the value of g_{m3} to be equal to $1/R_l$ (table 4.9). Its contribution to F adds up to that of R_l, resulting in a larger F_{min}.

The noise figure NF is plotted in figure 4.22 versus $g_{m2}R_S$ showing that:

- All the amplifiers are capable to achieve sub-3dB NF for proper values of $g_{m2}R_S$.
- Amplifiers e) and d) render the lowest F using the least number of devices: 3 (in this context, the differential pair is assumed as a compound device with an equivalent transconductance; matching device is not included). Amplifier a), g) and f) use 3 devices too. However, their NF is somewhat larger than that of e) and d) because the matching amplifier load device is a MOST for amplifier a) and g). For amplifier f) the contribution of R_l to F is larger because, for the same A_{VF}, the value of R_l is larger compared to the load resistance of the other amplifiers (i.e. larger output noise, see chapter 3). Amplifier c) suffers from a somewhat larger NF compared to amplifier d) because in the former the differential pair is replaced by a transconductor with 2 extra devices contributing noise. Amplifier h) shows the second largest NF. It suffers from the same problem as amplifier f) and, in addition, its uses 5 devices. Amplifier b) has the largest NF because the noise current of MOST g_{m3} adds to that of the matching amplifier load resistor (table 4.9). Next, it also uses 4 devices.

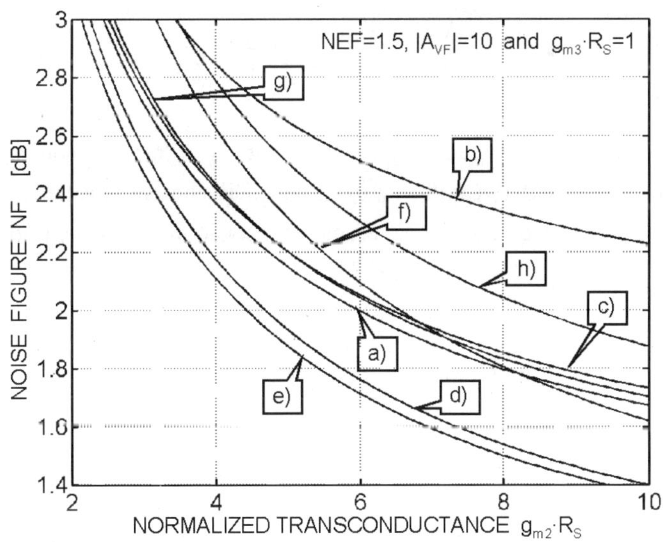

Figure 4.22: NF versus $g_{m2}R_S$ for $Z_{IN}=R_S$ and constant A_{VF}.

4.7.2 Noise factor versus power consumption

The noise factor of different amplifiers is compared assuming power consumption (and not $g_{m2}R_S$) as independent variable. This is because:

- Low power is an important requirement in many systems. In battery-operated systems, low power preserves battery lifetime. Next, it enables the use of a low-cost IC package.
- Generally, wide-band LNAs provide a lower F at larger power levels. Fixing the power budget, topologies that are inherently capable of lower F are then highlighted.

Neglecting the small contribution of the biasing circuitry, the power consumption P of the amplifiers in figure 4.16 can be written as:

$$P = \frac{V_{DD}}{R_S} \cdot \sum_i g_{m,i} R_S \cdot \left(\frac{I_i}{g_{m,i}} \right) \tag{4.10}$$

Where g_{mi} and $g_{m,i}/I_i$ are the transconductance and the g_m-efficiency of the transconductor implementing the VCCS (e.g.: MOST differential pair) and V_{DD} is the supply voltage. In equation (4.10), the sum is extended to the VCCSs determining the power of the amplifiers in figure 4.16 (i.e. g_{m1}, g_{m2} and eventually g_{m3}). Equation (4.10) can be written as:

$$P = \frac{V_{DD}}{R_S} \cdot \sum_i \frac{g_{m,i} R_S}{\xi_i} \cdot \left(\frac{I_{D,i}}{g_{m,i}} \right)_{MOST,CS} \tag{4.11}$$

The efficiency factor, $\xi_i > 0$, is used to relate the efficiency of a transconductor to that of a common-source (CS) MOST, $(g_{m,i}/I_{D,i})_{MOST,CS}$, which is chosen as reference. Assuming equal $(g_{m,i}/I_{D,i})_{MOST,CS}$ (i.e. optimal power efficiency), equation (4.11) yields to:

$$\begin{aligned} P &= P_{MOST,CS} \cdot \eta_{LNA} \\ P_{MOST,CS} &= \frac{V_{DD}}{R_S} \left(\frac{I_D}{g_m} \right)_{MOST,CS} \\ \eta_{LNA} &= \sum_i \frac{g_{m,i} R_S}{\xi_i} \end{aligned} \tag{4.12}$$

Equation (4.12) shows that the power consumption of the noise-cancelling amplifiers is the product of the (reference) power consumption $P_{MOST,CS}$ of a CS MOST with $g_m = 1/R_S$ and given g_m/I_D (as the other VCCS circuits in the amplifier) multiplied by the normalised power factor η_{LNA}(=$P/P_{MOST,CS}$). The latter circuit-dependent through the sum of $g_{m,i}R_S$ and the efficiency factor ξ_i. Table 4.10 shows the expressions of the normalised power factor η_{LNA} as a function of ξ_i and $g_{m2}\cdot R_S$ for the amplifiers in figure 4.16. Figure 4.23 shows NF for $Z_{IN}=R_S$, fixed gain A_{VF} versus $\eta_{LNA}=P/P_{MOST,CS}$.

LNA	ξ_1	ξ_2	ξ_3	$\eta_{LNA}=P/P_{MOS,CS}$	$g_{m2}R_S$	Condition
a	1+δ	1/4	-	$1/\xi_1+g_{m2}R_S/\xi_2$	$\xi_2(\eta_{LNA}-1/\xi_1)$	-
b	1+δ	1	1	$1/\xi_1+g_{m2}R_S/\xi_2+1/\xi_3$	$\xi_2(\eta_{LNA}-1/\xi_1-2/(A_{VF}\xi_3))$	$\lvert A_2\rvert=A_{VF}/2$
c	1+δ	1	1	$1/\xi_1+g_{m2}R_S/\xi_2+g_{m3}R_S/\xi_3$	$\xi_2(\eta_{LNA}-1/\xi_1-1/\xi_3)$	$\lvert A_2\rvert=A_{VF}/2$, $g_{m3}R_S=1$
d	1+δ	1/4	-	$1/\xi_1+g_{m2}R_S/\xi_2$	$\xi_2(\eta_{LNA}-1/\xi_1)$	-
e	1+δ	1	-	$1/\xi_1+g_{m2}R_S/\xi_2$	$\xi_2(\eta_{LNA}-1/\xi_1)$	-
f	1	1	-	$1/\xi_1+g_{m2}R_S/\xi_2$	$\xi_2(\eta_{LNA}-1/\xi_1)$	-
g	1	1	-	$1/\xi_1+g_{m2}R_S/\xi_2$	$\xi_2(\eta_{LNA}-1/\xi_1)$	-
h	1	1	1	$1/\xi_1+g_{m2}R_S/\xi_2+g_{m3}R_S/\xi_3$	$\xi_2(\eta_{LNA}-1/\xi_1-1/\xi_3)$	$\lvert A_3\rvert=1$, $g_{m3}R_S=1$

Table 4.10: Efficiency factor ξ and normalised power η_{LNA} (δ is the so-called MOST slope factor).

From this figure, we observe that:

- LNAs a) and d) provide the largest NF because they use a differential pair, which is rather power inefficient: $(g_m/I)_{PAIR}=(1/4)\cdot(g_m/I_D)_{MOS,CS}$.
- LNA e) offers the lowest NF due to the inherent efficiency of the single MOSFET (Analogously for LNA f and g). Despite its complexity, LNA c) has a far better NF than that of LNA d). Again, this is because the cascade of the CS stage, g_{m2}-R_2, with common-source MOSFET g_{m3} is more power efficient than the differential pair.

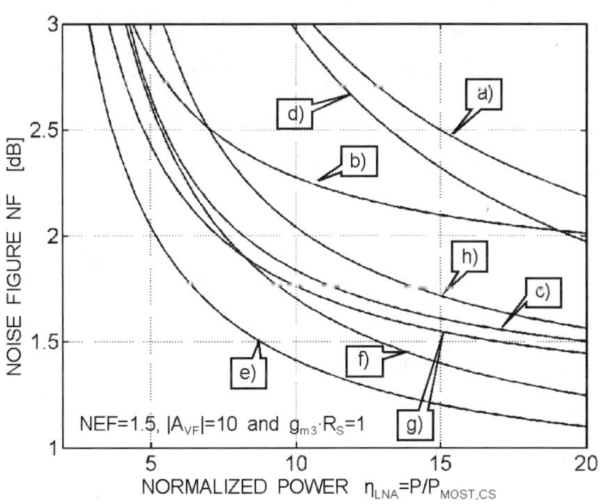

Figure 4.23: NF versus η_{LNA} for $Z_{IN}=R_S$ and constant A_{VF} (δ=0.3).

We now look at ways to enhance the efficiency factor ξ of a transconductor. This enables the same values of NF in figure 4.23 using less power or a lower NF is achieved for less power. Figure 4.24 shows some transconductor circuits providing a larger ξ.

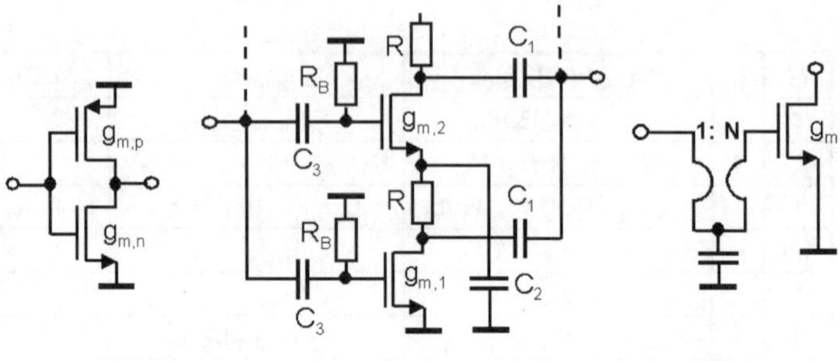

Figure 4.24: Transconductor circuits with improved efficiency factor ξ by exploiting: a)-b) MOST bias current re-use and c) Wide-band 1:N step-up transformer.

Circuits a) and b) achieve a larger ξ *re-using* the bias-current of another MOST. The g_m/I of the inverter in figure 4.24a is:

$$\frac{g_m}{I} = \frac{g_{m,n}}{I_D} \cdot \left(1 + \frac{g_{m,p}}{g_{m,n}}\right) \approx \frac{g_{m,n}}{I_D} \cdot \left(1 + \sqrt{\frac{K_p W_p}{K_n W_n}}\right) = \frac{g_{m,n}}{I_D} \cdot \xi \qquad (4.13)$$

where $K_{n(p)}=\mu_{n(p)}C_{ox,n(p)}$ and $L_p=L_n$ were assumed. Fixed $g_{m,n}/I_D$ and W_n (and so I_D), the inverter efficiency factor ξ is larger than 1. For $W_p=W_nK_n/K_p$, ξ is 2. This means that the g_m of an inverter is 2 times $g_{m,n}$ for the same bias I_D. For a typical CMOS process, K_n is about 2-3 times K_p. This requires large PMOST, which increases input capacitance C_{IN} as:

$$C_{IN} = C_{gs,n} + C_{gs,p} \approx C_{gs,n} \cdot \left(1 + \frac{W_p}{W_n}\right) \qquad (4.14)$$

and $C_{ox,n}=C_{ox,p}$ was used. For $W_p=W_n\mu_n/\mu_p$, the excess of input capacitance, $C_{gs,n}(\mu_n/\mu_p-1)$, can be substantial (e.g.: 2 or 3 times $C_{gs,n}$). Next, for a fixed bias current, the g_m/I of the inverter increases as the square root of W_p, while C_{IN} increases linearly with W_p. Thus, the inverter unity-gain cut-off frequency, $f_T=g_m/(2\pi C_{IN})$, drops as the inverse of the square root of W_p. To mitigate the previous problems the circuit in figure 4.24b may be used. Here, the bias current of the bottom NMOST is re-used by a MOST of the same type. The total g_m approaches then the sum of the g_m of the stacked MOSTs (i.e. ξ = number of stacked MOSTs). Nevertheless, this solution requires extra resistors, capacitors and dc sources to bias correctly the MOSTs and ground their source terminals. These components increase chip-area and introduce bandwidth limitations. Moreover, the output noise of R may be not negligible. These issues exacerbate at a low supply-voltage, due to the scarce voltage headroom available for the stacked MOSTs and R's. Figure 4.24c shows an alternative approach..A step-up 1:N transformer in front of a MOST boosts the g_m to $N \cdot g_m$ ($\xi = N$). Unfortunately, wide-band transformers of acceptable performance are difficult to integrate, especially in CMOS.

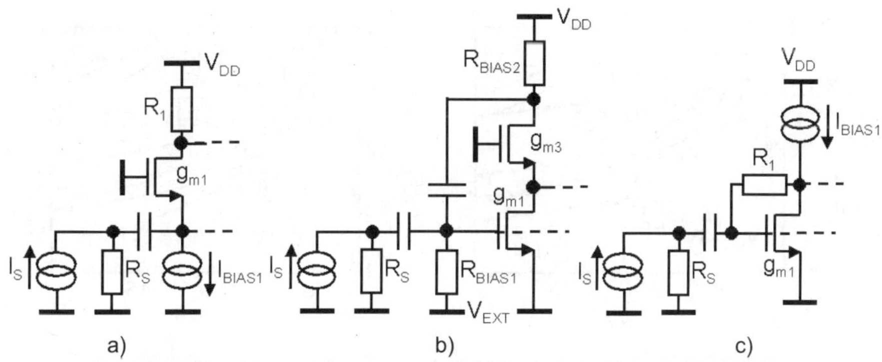

Figure 4.25: Biasing of the matching stages used in the LNAs of figure 4.16.

4.7.3 Noise factor versus power consumption with biasing noise

So far, we have considered only the noise generated by the devices in the signal path. However, noise from the biasing devices can degrade F especially at low supply-voltage. Figure 4.25 shows the (simplified) biasing for the matching stages used of the LNAs of figure 4.16. Here, the noise currents associated to I_{BIAS1}, R_{BIAS1} and R_{BIAS2} add to the input signal I_S. The biasing noise factor, F_{BIAS}, is:

$$F_{BIAS} = 1 + \frac{\overline{I^2_{n,BIAS1}}}{\overline{I^2_{n,RS}}} + \frac{\overline{I^2_{n,BIAS2}}}{\overline{I^2_{n,RS}}} = 1 + (G_{n,BIAS1} + G_{n,BIAS2}) \cdot R_S \qquad (4.15)$$

$G_{n,BIAS1} = NEF \cdot g_{m,BIAS1}$ (R_{BIAS1}: NEF=1 and $g_{m,BIAS1} = 1/R_{BIAS1}$) and $G_{n,BIAS2} = 1/R_{BIAS2}$ are noise conductance (figure 4.25a, $G_{n,BIAS2}=0$). F_{BIAS} is small for $(G_{n,BIAS2}+G_{n,BIAS2}) \cdot R_S << 1$. This requires a large voltage across I_{BIAS1} and R_{BIAS2}. For a fixed A_{VF} and supply voltage V_{DD}, F_{BIAS} can be non-negligible. For the LNA f) and h) the bias noise current, $I_{n,BIAS1}$, is injected into the drain node of the matching device. F_{BIAS1} is then smaller because the signal at this node is large (i.e. gain>1). Next, the output noise due to $I_{n,BIAS1}$ is cancelled too. Figure 4.26 shows $F+F_{BIAS}$ versus η_{LNA} for different values of the supply voltage V_{DD}. The expressions of the bias noise factor, $F_{BIAS}(A_{VF}, V_{DD}, V_{GT1}, \theta)$, are given in appendix B. From this figure, we can conclude that:

- The total NF of the LNAs a) to e) exploiting a common-gate matching stage and LNA g) degrade significantly as V_{DD} drops down to 1.5V. These LNAs are less suitable to operate at low supply voltage due to the contribution of F_{BIAS}.
- LNAs f) to h) are unaffected by matching-stage biasing noise.

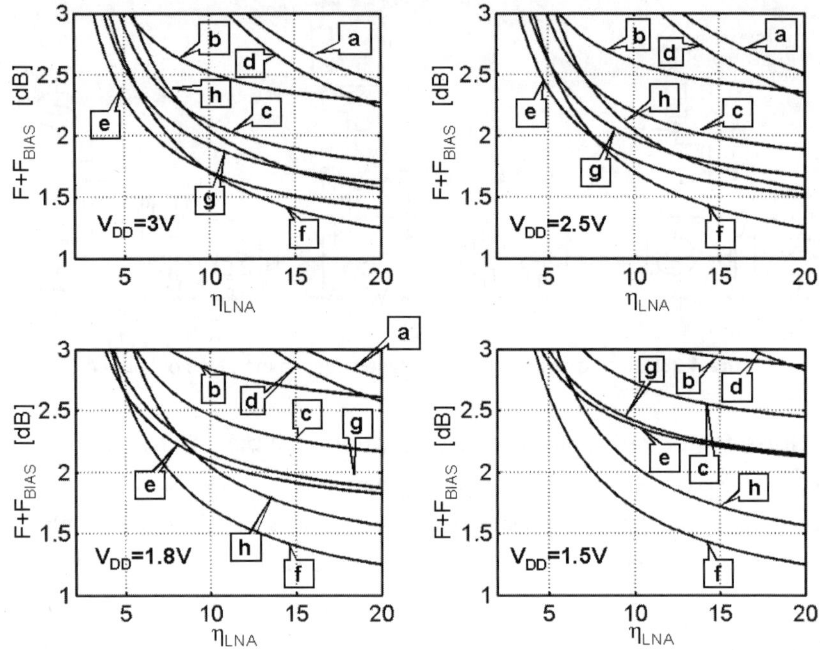

Figure 4.26: $F+F_{BIAS}$ versus η_{LNA} for different values of supply voltage V_{DD} ($|A_{VF}|=10$, NEF=1.5, $V_{gs1}-V_{T0}=0.2V$, $\theta=0.8$ 1/V and $\delta=0.3$ have been assumed).

Conclusions: From the previous comparison of the noise cancelling LNAs in figure 4.16 it can be concluded that LNA f) offers potentially the best noise performance as far as $F+F_{BIAS}$ versus the normalised power η_{LNA} for $Z_{IN}=R_S$ and fixed A_{VF} is concerned. Another advantage of this LNA is the lower output impedance of the shunt-feedback CS (CSSF) stage compared to the CG amplifier stage. As such, LNA f) can have superior high-frequency behaviour provided the capacitance at the output node of the CSSF stage does not differ significantly from the capacitance at the output of the CG amplifier stage.

4.8 Noise Cancelling Properties

This paragraph discusses important properties of noise cancelling such as:

1. Robustness.
2. Simultaneous noise and power matching.
3. Distortion cancelling.

Although these properties concern all the LNAs in figure 4.16, for sake of convenience, we will often refer to the LNA in figure 4.16f or the amplifier model in figure 4.20c.

4.8.1 Robustness

In figure 4.27, the transfer function of the signal and the matching device noise current $I_{n,i}$ are again showed for a generic noise-cancelling LNA.

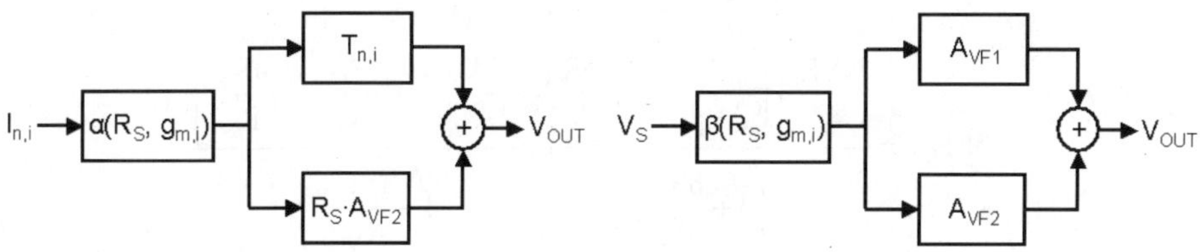

Figure 4.27: Feed-forward representation of the noise and signal transfer function.

According to this figure, noise cancelling occurs for $T_{n,i}+R_S{\cdot}A_{VF2}=0$. This requires the module of the noise transfers $T_{n,i}$ and $R_S{\cdot}A_{VF}$ to be equal. In practice, due to device parameter variations (e.g.: process spread), this identity can be realised to a limited extent. Nevertheless, noise cancelling is robust to these variations.

- The noise cancelling equation, $T_{n,i}+R_S{\cdot}A_{VF2}=0$, depends on a *reduced* set of circuit parameters. Consider, for instance, the LNA in figure 4.16f, which is redrawn in figure 4.28. The impedance Z_Y and Z_L (and the g_m of the matching device) do *not* affect the solution of $T_{n,i}+R_S{\cdot}A_{VF2}=0$ (i.e. $1+R/R_S-g_{m2}/g_{m3}=0$). This is because Z_Y and Z_L affect the two feed-forward paths in figure 4.27 in the same fashion.

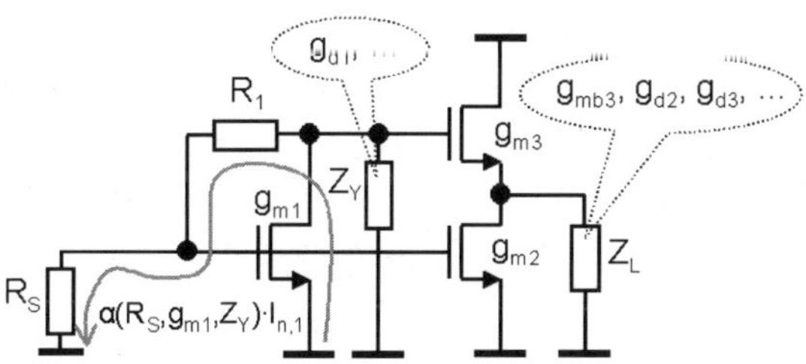

Figure 4.28: Effect of Z_Y and Z_L on noise cancelling for the amplifier in figure 4.16f.

- The performance of the noise cancellation technique is only modestly sensitive to device parameter variations (e.g.: process-spread). Consider the variation of the source impedance and gain of the voltage-sensing amplifier, $R_S+\delta R_S$ and $A_{VF2}+\delta A_{VF2}$. The variation of EF_{MD} from its nominal value for $Z_{IN}=R_S$ and $T_{n,i}+R_S\cdot A_{VF2}=0$ is:

$$\delta EF_{MD}=\frac{NEF\left[\delta T_{n,i}+T_{n,i}+(R_S+\delta R_S)\cdot(A_{VF2}+\delta A_{VF2})\right]^2}{R_S(R_S+\delta R_S)\cdot(A_{VF2}+\delta A_{VF2}+A_{VF1})^2}$$

where $\delta T_{n,i}$ is the variation of $T_{n,i}$ caused by δR_S. Table 4.8 shows the relation among A_{VF1}, A_{VF2} and $T_{n,i}$ for the amplifiers in figure 4.16. For amplifier 4.16f, for instance, $A_{VF1}=A_{VF2}+2$, $T_{n,i}=-R_SA_{VF2}$ and $\delta T_{n,i}=\delta R_S$. For this case, one can write:

$$\delta EF_{MD}=\frac{\frac{NEF}{4}\cdot\left[\left(\frac{1}{A_{VF2}}+1\right)\cdot\frac{\delta R_S}{R_S}+\frac{\delta A_{VF2}}{A_{VF2}}+\frac{\delta A_{VF2}}{A_{VF2}}\cdot\frac{\delta R_S}{R_S}\right]^2}{\left(1+\frac{\delta R_S}{R_S}\right)\cdot\left(\frac{1}{2}\cdot\frac{\delta A_{VF2}}{A_{VF2}}+1+\frac{1}{A_{VF2}}\right)^2}$$

Contours of δEF_{MD} are shown in figure 4.29 for NEF=1.5 and $A_{VF2}=-7$. For a $\delta R_S/R_S$ and $\delta A_{VF2}/A_{VF2}$ as large as ±20%, δEF_{MD} rises to only 0.1, ten times smaller than the contribution of the source. For $\delta R_S/R_S\approx 0$, e.g. when R_S is the impedance seen "looking into" a properly terminated cable, the gain error $|\delta A_{VF2}/A_{VF2}|$ can be as large as ±36%.

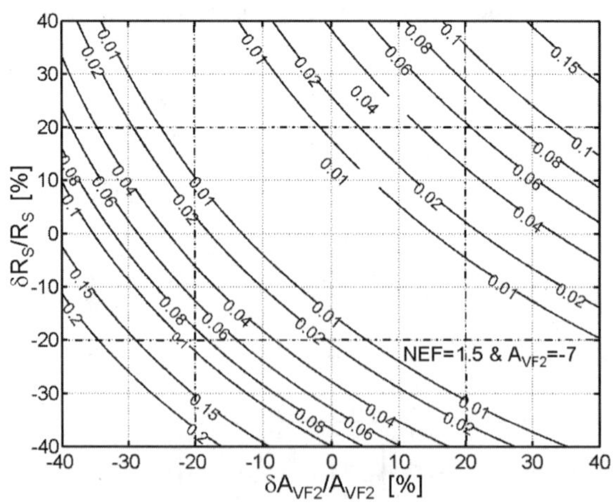

Figure 4.29: Contour of δEF_{MD} vs. $\delta A_{VF2}/A_{VF2}$ and $\delta R_S/R_S$.

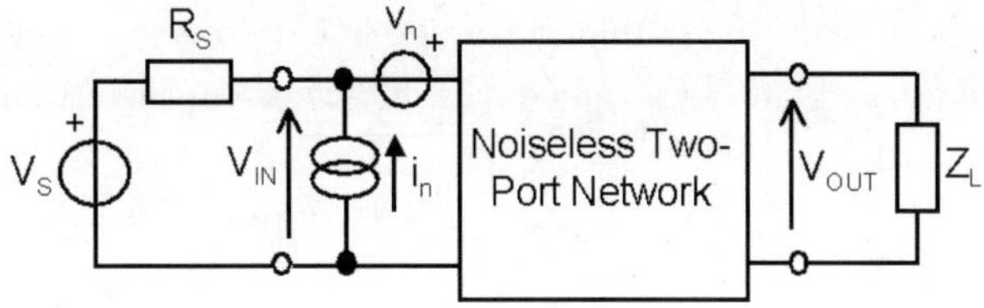

Figure 4.30: Two-port noise model: "i_n" and "v_n" are the equivalent input noise current and voltage.

4.8.2 Simultaneous noise and power matching

The noise factor of a two-port circuit in figure 4.30 can be written as [16]:

$$F(G_S) = 1 + \frac{\overline{v_n^2}\cdot(G_S+G_C)^2 + \overline{i_{n,U}^2}}{\overline{I_{n,S}^2}} = F_{MIN} + \frac{R_n}{G_S}\cdot(G_S - G_{S,OPT})^2$$

$$F_{MIN} = 1 + 2\cdot R_n\cdot(G_S + G_{S,OPT}) \quad R_n = \frac{\overline{v_n^2}}{4kT\Delta f} \quad G_S = \frac{\overline{I_{n,S}^2}}{4kT\Delta f} \quad G_U = \frac{\overline{i_{n,U}^2}}{4kT\Delta f}$$

$$G_{S,OPT} = \sqrt{G_C^2 + \frac{G_U}{R_n}}; \quad G_C = \frac{\overline{v_n\cdot i_n}}{\overline{v_n^2}} \quad \text{and} \quad G_U = \frac{\overline{(i_n - G_C\cdot v_n)^2}}{4kT\Delta f} \tag{4.16}$$

where the above noise quantities have the usual meaning as in [16]. Relations (4.16) show the existence of an optimum value of the source conductance, $G_{S,OPT}$, which renders the minimum noise factor, F_{MIN} ($G_S=1/R_S$ is the source conductance). If $G_{S,OPT}$ is equal to G_{IN}, the input of the two-port is simultaneously optimised for noise and power transfer when G_{IN} is equal to G_S. In this case, we say that the two-port input is simultaneously noise-power matched (i.e. maximum transfer of power to the input and $F=F_{MIN}$). In general, $G_{S,OPT}$ differs from G_{IN} and one can either match the two-port input for noise or for power. Noise-power matching can be achieved using a wide-band matching network, which transforms G_S into $G_{S,OPT}$. This network can be rather complex, consisting of several area-consuming capacitors and inductors. In this paragraph, it will be shown that LNAs exploiting noise cancelling can be designed for simultaneous noise and power matching without an extra wide-band matching network[XIV].

The optimum source conductance, $G_{S,OPT}$, of a two-port with M independent noise sources, $I_{n,k}$, can be written as (see appendix C):

[XIV] This is strictly true only when a MOST can be modelled as a noisy VCCS.

$$G_{S,OPT} = G_{IN} \cdot \frac{\sum_{k=1}^{M} I_{n,k} \cdot H_{n,open,k}}{\sum_{k=1}^{M} I_{n,k} \cdot H_{n,short,k}}$$

$$H_{n,short,k} = \left. \frac{V_{OUT}}{I_{n,k}} \right|_{V_{IN}=0} \quad \text{and} \quad H_{n,open,k} = \left. \frac{V_{OUT}}{I_{n,k}} \right|_{I_{IN}=0} \tag{4.17}$$

$H_{n,open(short),k}$ is the transfer from $I_{n,k}$ to the output of the two-port upon a opened (shorted) input port. From (4.17), simultaneous noise and power matching requires:

$$\frac{\sum_{k=1}^{M} I_{n,k} \cdot H_{n,open,k}}{\sum_{k=1}^{M} I_{n,k} \cdot H_{n,short,k}} = 1 \ \forall k \ \Leftrightarrow \ \sum_{k=1}^{M} I_{n,k} \cdot (H_{n,open,k} - H_{n,short,k}) = 0 \ \forall k \tag{4.18}$$

Finally, equation (4.18) yields the condition:

$$H_{n,open,k} = H_{n,short,k} \quad \forall\ k \tag{4.19}$$

From relation (4.19), noise and power matching requires that, for each noise source of the two-port, the noise transfer $H_{n,k}$ does not differ when passing from a shorted to opened input port. Let's now consider two examples: the negative feedback LNA in figure 4.6a and the noise-cancelling LNA in figure 4.16f. Table 4.11 shows their $H_{n,open}$ and $H_{n,short}$.

LNA	$H_{n,open}$	$H_{n,short}$
Fig. 4.6a	Ri: $R_iAv/(1-Av)$	Ri: 0
	"Av": $R/(1-Av)$	"Av": R
Fig. 4.16f	R: -R	R: -R
	M1: $(Av -1)/g_{m1}$ Upon cancellation $R/(g_{m1} \cdot R_S)=R$	M1: R

Table 4.11: Noise transfers $H_{n,short}$ and $H_{n,open}$.

For the feedback LNA, the noise transfer of the feedback resistor and the loop amplifier "Av" (Assuming a CS stage plus an ideal buffer) differ upon open and shorted input (table 4.11). This LNA is then not capable of simultaneous noise and power matching. For the noise cancelling LNA, the matching device noise transfers are identical upon $g_{m1} \cdot R_S=1$ and noise cancellation (i.e. $Av=-1-R/R_S$), so simultaneous noise and power match is possible without any matching network.

Figure 4.31 shows the plot of NF and NF_{MIN} versus R_S for the 50Ω input LNA in figure 4.20c (i.e. replacing amplifier Av with an ideal voltage controlled voltage source). As expected, NF is equal to NF_{MIN} only for $R_S=R_{IN}=50\Omega$. Figure 4.31 shows the same noise figure plots (i.e. NF_{20} and $NF_{MIN,20}$) for amplifier "Av" with an equivalent input noise resistance $R_{n,EQ,IN}=20\Omega$. Clearly, the simultaneous noise and power matching is not affected by the noise of "Av".

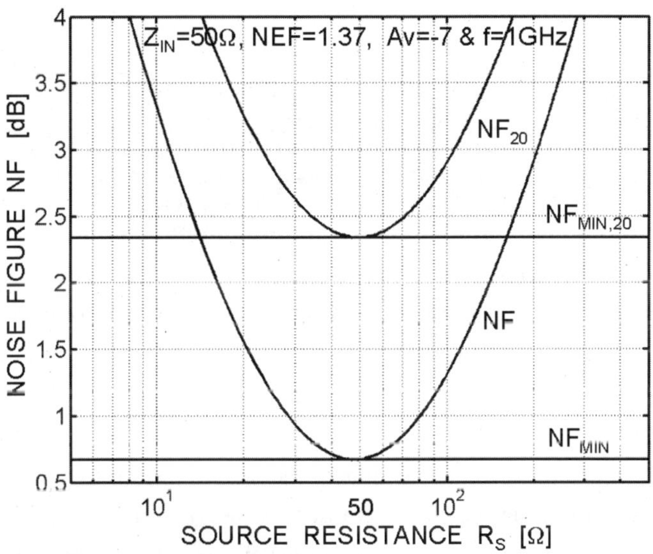

Figure 4.31: Simulated NF (NF_{20}) and NF_{MIN} ($NF_{MIN,20}$) versus R_S for the LNA in figure 4.20c (MOS model 9) for a noiseless (noisy) amplifier "Av". The matching-stage provides $R_{IN}\approx 50\Omega$ for $V_{GS}-V_{TO}\approx 250mV$. R=300Ω renders a voltage gain ($=V_Y/V_X$) of 12.8dB for the matching stage.

Figure 4.32 shows NF and NF_{MIN} versus R_S for the 50Ω input LNA in figure 4.32-II where the loop amplifier is represented by an ideal voltage controlled voltage source with an equivalent input noise resistance $R_{n,EQ,IN}=30\Omega$. As expected, $NF(R_S=50\Omega)=2.851dB$ and $NF_{MIN}(R_{S,OPT}=97\Omega)=2.514dB$ are different so this feedback amplifier is not capable of simultaneous noise and power matching.

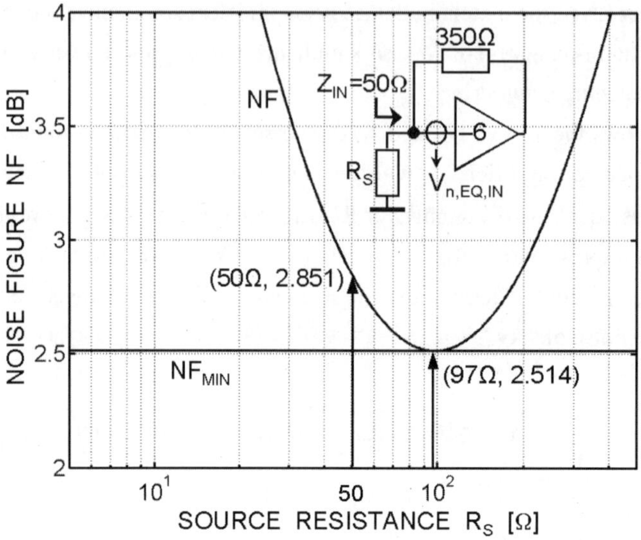

Figure 4.32: Simulated NF and NF_{MIN} versus R_S of the LNA in figure 4.7-III for a noisy amplifier "Av" with $R_{n,EQ,IN}$=30Ω.

4.8.3 Distortion cancelling

In this section, it will be shown that the same mechanism allowing for the cancellation of the output noise of the matching device can be also exploited to cancel the distortion components produced by the matching device. It will be assumed that the non-linear voltage-to-current conversion (V_{GS}-I_{DS}) of the MOSFET is the only cause of distortion.

If we use a Taylor approximation to model the weakly nonlinear behavior, the drain current of the matching device can be written as $I_D=g_{mi}V_X+I_{NL}$, where I_{NL} denotes all nonlinear high order terms. From inspection of the circuit in figure 20c, the signal voltage at nodes X and Y can now be written as:

$$\begin{aligned} V_X &= V_S - R_S(g_{mi}V_X + I_{NL}) \\ V_Y &= V_S - (R_S + R)(g_{mi}V_X + I_{NL}) \end{aligned} \tag{4.20}$$

After some manipulations one obtains:

$$\begin{aligned} V_X &= \frac{V_S}{1+g_{mi}R_S} - \frac{R_S I_{NL}}{1+g_{mi}R_S} \\ V_Y &= V_S \cdot \frac{1-g_{mi}R}{1+g_{mi}R_S} - \left(1+\frac{R}{R_S}\right) \cdot \frac{R_S I_{NL}}{1+g_{mi}R_S} \end{aligned} \tag{4.21}$$

Equation (4.21) shows that the distortion voltage at node Y, $-(R_S+R)\cdot I_{NL}/(1+g_{mi}R_S)$, has $1+R/R_S$ times higher amplitudes then at node X, $-R_S\cdot I_{NL}/(1+g_{mi}R_S)$, and has *equal sign.* Therefore, a gain $Av=-(1+R/R_S)$ cancels the output noise as well as all nonlinear terms contributed by the matching device. This gain is the same value required to cancel the output noise due to the matching device, thereby allowing for *simultaneous* noise and distortion cancellation. Figure 4.33 highlights the distortion cancellation in the frequency-domain, for the relevant case of 3rd order IM distortion. Two equal carriers at frequencies f_1 and f_2 produce close-in 3rd order IM distortion products at frequency $2f_1$-f_2 and $2f_2$-f_1. These undesired 3rd order products are in anti-phase at nodes X and Y, while the (desired) signal terms are in phase, explaining while distortion products cancel at the output.

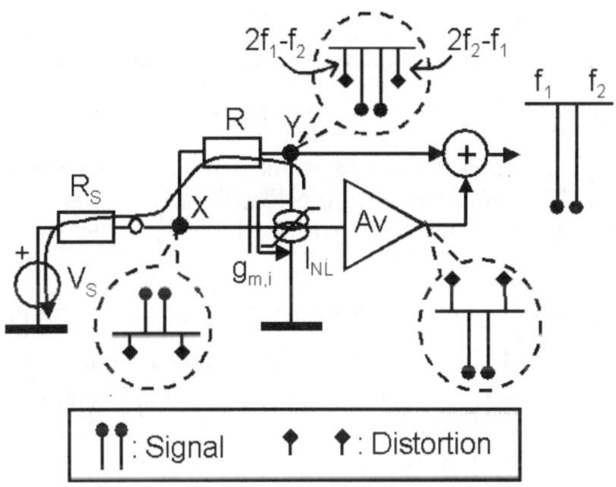

Figure 4.33: Cancellation of 3rd order intermodulation distortion terms sketched in the frequency domain.

Figures 4.34 shows the simulated (using PSS-analysis) input-referred 2nd and 3rd order IM intercept points IIP2 and IIP3 (p_{in}=-30dBm@50Ω for each tone) versus |Av| for the LNA in figure 4.33. Amplifier "Av" is replaced by an ideal voltage controlled voltage source. IIP2 and IIP3 peak for $|Av|= 1+R/R_S=1+300/50=7$ where the distortion cancellation occurs. Furthermore, the IIPs are always larger than for Av=0 (i.e. the IIPs of the matching stage standalone). Note that the cancellation occurs despite the effect of the non-linear output conductance of the MOSFET.

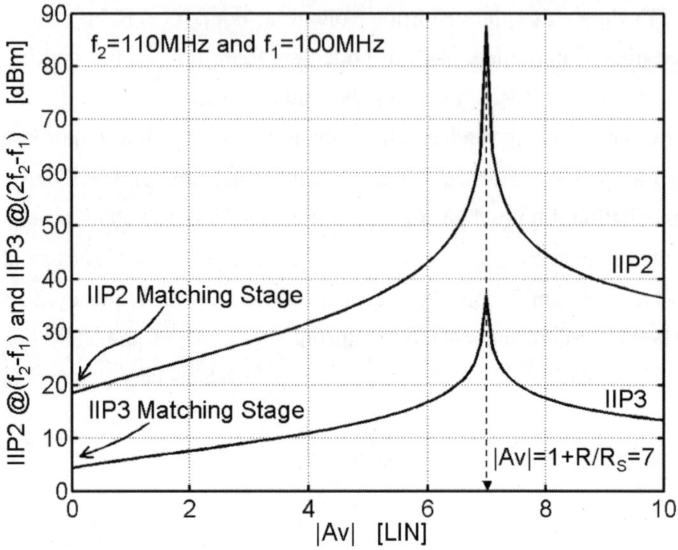

Figure 4.34: Simulated IIP2 and IIP3 versus |Av| for the amplifier in figure 4.32 (MOS model 9). The matching-stage provides $R_{IN} \approx R_S = 50\Omega$ for $V_{GS}\text{-}V_{T0} \approx 250$mV while $R=300\Omega$ renders a gain (V_Y/V_X) of 12.8dB for the matching stage.

The distortion cancellation described so far does not take into account the non-linearity of the voltage amplifier "Av". This can be done, for instance, approximating its input-output characteristic by a 3rd order Taylor function with coefficients Av, Av,2 and Av,3. For this case, using the method of the direct-currents [17], the 2nd and 3rd order IM distortion at the output of the amplifier in figure 4.33 can be written as:

$$\mathrm{IMD2} = \frac{V_S}{(1+g_{m1}R_S)^2} \cdot \left| \frac{Av{,}2 \cdot (1+g_{m1}R_S) - g_2 R_S \left(Av + 1 + \frac{R}{R_S} \right)}{1 - g_{m1}R + Av} \right|$$

$$\mathrm{IMD3} = \frac{V_S^2 \left(\frac{3}{4} \right)}{(1+g_{m1}R_S)^3} \cdot \left| \frac{Av{,}3 \cdot (1+g_{m1}R_S) - 2g_2 R_S Av{,}2 + \left(\frac{2g_2^2 R_S^2}{1+g_{m1}R_S} - g_3 R_S \right) \left(Av + 1 + \frac{R}{R_S} \right)}{1 - g_{m1}R + Av} \right| \tag{4.22}$$

$g_{mi}=\partial I_{DS}/\partial V_{GS}$, $g_2=(1/2)\partial^2 I_{DS}/\partial V_{GS}^2$ and $g_3=(1/6)\partial^3 I_{DS}/\partial V_{GS}^3$ are the Taylor coefficients of the matching device and $g_2>0$ and $g_3<0$ holds for a MOSFET biased in strong-inversion and saturation. From equation (4.22), two distortion contributions can be isolated at the output of the LNA. The output distortion when "Av" is linear: $-g_2R_S(Av+1+R/R_S)$ for IMD2 and $(2(g_2R_S)^2/(1+g_{m1}R_S)-g_3R_S)(Av+1+R/R_S)$ for IMD3. The distortion caused by the non-linearity of "Av": $Av,2(1+g_{m1}R_S)$ for IMD2 and $Av,3(1+g_{m1}R_S)-2g_2R_SAv,2$ for IMD3. The intercept points V_{IIP2} and V_{IIP3} are then equal to:

$$V_{IIP2} = \frac{(1+g_{m1}R_S)^2|1-g_{m1}R+Av|}{\left|Av,2\cdot(1+g_{m1}R_S)-g_2R_S\left(Av+1+\frac{R}{R_S}\right)\right|}$$

$$V_{IIP3}^2 = \frac{\left(\frac{4}{3}\right)\cdot\frac{(1+g_{m1}R_S)^3}{|1-g_{m1}R+Av|}}{\left|Av,3\cdot(1+g_{m1}R_S)-2g_2R_SAv,2+\left(\frac{2g_2^2R_S^2}{1+g_{m1}R_S}-g_3R_S\right)\left(Av+1+\frac{R}{R_S}\right)\right|} \tag{4.23}$$

From (4.22) and (4.23), the cancellation conditions for $Z_{IN}=R_S$ are respectively:

$$\frac{R}{R_S} = -1 - Av + 2\frac{Av,2}{g_2R_S} \qquad \frac{R}{R_S} = -1 - Av + 2\cdot\frac{g_2R_SAv,2-Av,3}{g_2^2R_S^2-g_3R_S} \tag{4.24}$$

where $R/R_S>1$ (i.e. negative gain for the matching stage) and $Av<0$ must hold. Equation (4.24) renders the value of the ratio R/R_S that allows for the output distortion cancellation given R_S and the characteristics of amplifier "Av" and the matching device.

From equations (4.22), the following possible scenarios regarding the effects of the non-linearity of "Av" on the distortion canceling are identified:

- The output distortion caused by the non-linearity of amplifier "Av" adds *in-phase* to the distortion generated by the matching stage: $Av,2>0$ and $Av,3(1+g_{m1}R_S)-2g_2R_SAv,2<0$. In this case, cancellation of the total distortion at the output of the LNA is achieved by *increasing* the strength of the distortion components at node Y. This requires a ratio R/R_S that is *larger* than $1-Av$.
- The output distortion caused by the non-linearity of amplifier "Av" adds 180° *out-of-phase* to the distortion generated by the matching stage: $Av,2<0$ and $Av,3(1+g_{m1}R_S)-$

$2g_2R_SA_{v,2}>0$. In this case, two situations can be distinguished. The distortion component caused by "Av" is *lower* than that caused by the matching-stage: $A_{v,2}(1+g_{m1}R_S)-g_2R_SA_v>0$ for IMD2 and $A_{v,3}(1+g_{m1}R_S)-2g_2R_SA_{v,2}+A_v(2(g_2R_S)^2/(1+g_{m1}R_S)-g_3R_S)<0$ for IMD3. The distortion component caused by "Av" is *larger* than that caused by the matching stage: $A_{v,2}(1+g_{m1}R_S)-g_2R_SA_v<0$ for IMD2 and $A_{v,3}(1+g_{m1}R_S)-2g_2R_SA_{v,2}+A_v(2(g_2R_S)^2/(1+g_{m1}R_S)-g_3R_S)>0$ for IMD3. In the former case, exact distortion cancellation is still possible but this time by *decreasing* the strength of the distortion components at node Y. This requires a ratio R/R_S *smaller* than $1-A_v$. In the last case, distortion canceling will *not* occur because the necessary sign relation among the distortion components is irreparably altered.

- The value of R/R_S needed to cancel the 2nd and the 3rd order distortion is not the same.
- Distortion canceling occurs for a ratio R/R_S that is not the one required to cancel noise. In other words, *simultaneous noise-distortion cancellation* is hampered. In this respect, two measures can be taken: (a) to choose for noise or distortion cancellation thereby accepting the degradation of the other and (b) to minimize the 2nd and 3rd order Taylor coefficients of "Av" such that distortion and noise cancellation are approximately achieved both. However, at high frequencies, it is difficult to design amplifier "Av" such that it produces much less output distortion than that generated by matching stage. This is an important difference with respect to noise canceling where the contribution to F of "Av" is easily made smaller than that of the matching device by increasing the g_m of its input stage. Moreover, noise and distortion cancellation may lead to contradictory requirements. For instance, a large g_m (i.e. low F) at acceptable power levels demand for a relatively large g_m/I_D. On the other hand, the linearity of the MOST improves when the g_m/I_D is low (i.e. large $V_{GS}-V_{T0}$).

Figure 4.35 and 4.36 show the plots of the simulated IIP2 and IIP3 (p_{in}=-30dBm@50Ω for each tone) versus the ratio R/R_S as obtained by a two-tone test (PSS analysis). According to the above analysis, cancellation of the total 2nd and 3rd order distortion at the output of the amplifier in figure 4.33 occurs for a proper ratio R/R_S approximately indicated by equation (4.24).

As final note, it is worth wise to mention that, in contrast to noise cancellation, distortion cancellation is *signal-dependent*. This is a shortcoming especially for wide-band systems where the distortion requirements are specified for signal levels that are generally larger than that used for narrow-band systems.

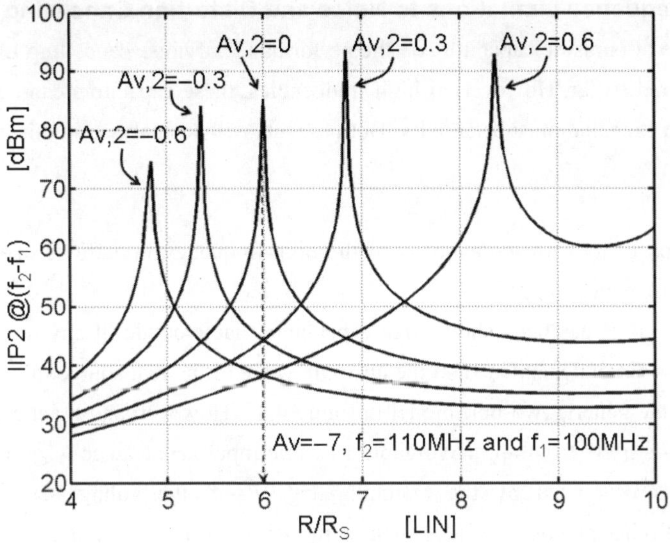

Figure 4.35: Simulated IIP2 versus R/R_S for different values of Av,2 (MOS model 9) for the amplifier in figure 4.33. The matching-stage provides $R_{IN} \approx R_S = 50\Omega$ for $V_{GS}-V_{T0} \approx 250mV$ while $R=300\Omega$ renders a voltage gain ($=V_Y/V_X$) of 12.8dB for the matching stage.

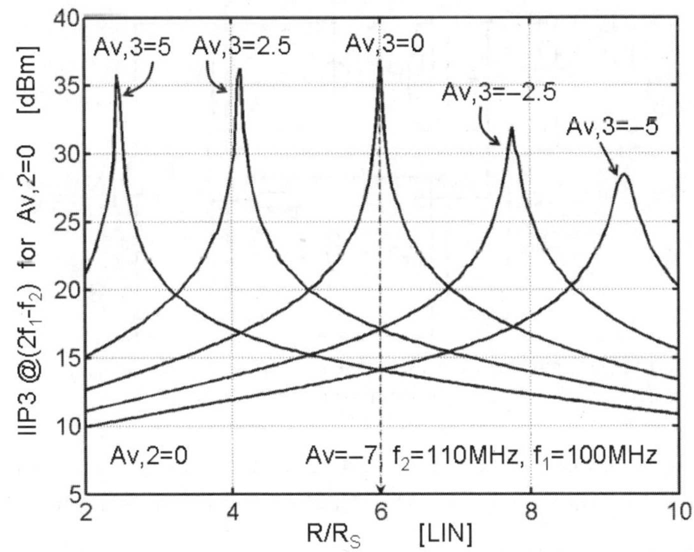

Figure 4.36: Simulated IIP3 versus R/R_S for different values of Av,3 and Av,2=0 (MOS model 9) for the amplifier in figure 4.33. The matching-stage provides $R_{IN} \approx R_S = 50\Omega$ for $V_{GS}-V_{T0} \approx 250mV$ while $R=300\Omega$ renders a voltage gain ($=V_Y/V_X$) of 12.8dB for the matching stage.

4.9 High-Frequency Limitations to Noise and Distortion Cancelling

The influence of parasitic capacitors on the performance of noise cancelling LNAs has not been considered so far. However, at high frequencies, these capacitors can degrade noise and distortion cancellation substantially. Both aspects will be analysed in this paragraph.

4.9.1 Noise

This paragraph deals with the analysis of the high-frequency limitations of LNA circuit exploiting noise cancelling.

In paragraph 4.6, it has been shown that a peculiar characteristic of any noise canceling amplifier is its direct dependence on the absolute value of the real source impedance R_S as shown again the generic two-port model in figure 4.37. However, any practical realization of figure 4.37 suffers from an unavoidable parasitic input capacitance to ground C_1. This capacitance is originated by the matching-stage itself, the voltage-sensing two-port amplifier and other practical devices such as the RF input bond-pad and the electrostatic protection device (ESD). Therefore, it can be substantial (e.g.: 1pF or more).

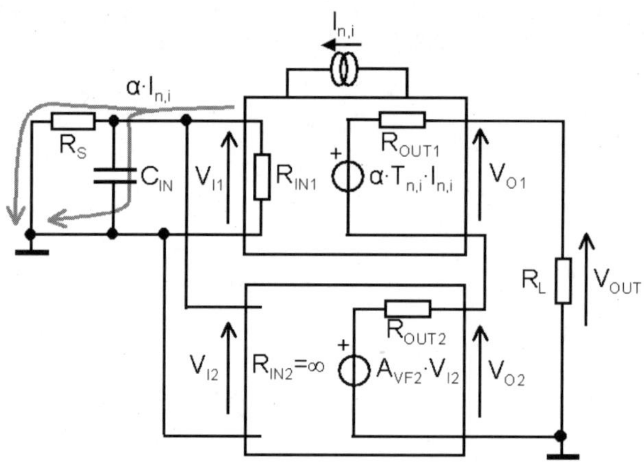

Figure 4.37: Two-port model of a noise-cancelling amplifier showing the effect of the input capacitance C_1 (V_S=0).

From the point of view of the noise current flowing out from the matching device, C_1 acts in parallel to the real source impedance R_S. This means that this noise current, $\alpha \cdot I_{n,i}$, sees an effective frequency-dependent source impedance, $Z_S(s)=R_S/(1+s \cdot R_S \cdot C_1)$. Per se, the frequency-dependence of $Z_S(s)$ is not the issue. The problem with $Z_S(s)$ is that it affects the two feed-forward noise transfers from $\alpha \cdot I_{n,i}$ to the output in a different fashion. In other words, the two noise paths experience different frequency-dependence leadings to a lower degree of cancellation as the frequency increases. Let's now analyse the effect of C_1 for the noise-cancelling amplifier in figure 4.38a.

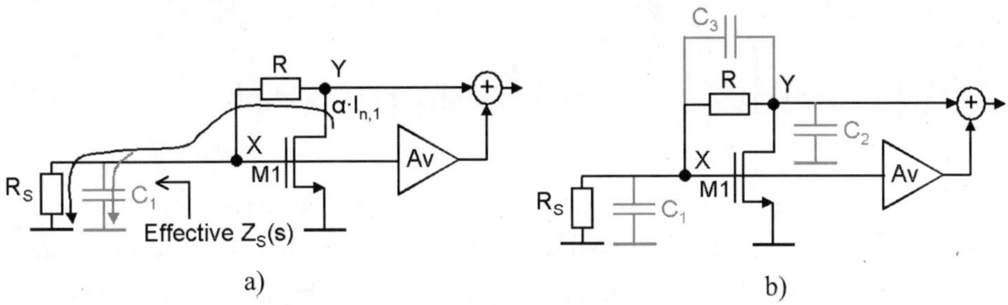

Figure 4.38: Noise cancelling LNA with input capacitance C_1 (a) and plus parasitic C_2 and C_3 (b).

The transfer from the matching device noise current, $H_{n,1}(s)= V_{OUT}/I_{n,1}$, is:

$$H_{n,1}(s)=H_{n,1,1}(s)+H_{n,1,2}(s)=\alpha(Z_S(s),g_{m1})\cdot[R_1+Z_S(s)+Z_S(s)\cdot Av]$$

$$Z_S(s)=\frac{R_S}{1+s\cdot R_S C_1} \quad \text{and} \quad \alpha(Z_S(s),g_{m1})=\frac{1+s\cdot R_S\cdot C_1}{1+\dfrac{s\cdot R_S\cdot C_1}{1+g_{m1}\cdot R_S}}\cdot\frac{1}{1+g_{m1}\cdot R_S} \tag{4.25}$$

where $H_{n,1,1}(s)$ and $H_{n,1,2}(s)$ are the contributions to $H_{n,1}(s)$ of the two feed-forward paths to the output. Equations (4.25) show that C_1 affects differently the two feed-forward noise transfers. This is most evident by looking at the values of $|H_{n,1,1}(f)|$ and $|H_{n,1,2}(f)|$ for $f\rightarrow\infty$. The former is R_1 and the latter is 0. For $-Av=R_1/R_S+1$, equation (4.25) yields to:

$$H_{n,1}(s)=\alpha(Z_S(s),g_{m1})\cdot R_1\cdot\left[1-\frac{Z_S(s)}{R_S}\right] \tag{4.26}$$

$$H_{n,1}(s)=\frac{s\cdot R_S C_1}{1+\dfrac{s\cdot R_S C_1}{1+g_{m1}R_S}}\cdot\frac{R_1}{1+g_{m1}R_S} \tag{4.27}$$

From another point of view, equation (4.27) shows the existence of a zero-frequency in the origin for $H_{n,1}(s)$, which is mainly responsible of the degradation of noise cancelling.
The excess noise factor of the matching device, $EF_{MD}(f)$, is equal to (for $g_{m1}\cdot R_S=1$):

$$EF_{MD}(f)=NEF\cdot\left(\frac{f}{f_0}\right)^2 \tag{4.28}$$

with $f_0=1/(\pi \cdot R_S \cdot C_1)$ the LNA input pole. Equation (4.28) shows that exact cancellation occurs only at "dc", with EF_{MD} rising with the square of the frequency. When amplifier "Av" is implemented as shown in figure 4.16f, the noise factor F(f) can be written as:

$$F(f)=1+(F_{DC}-1)\cdot\left[1+\left(\frac{f}{f_0}\right)^2\right]+NEF\cdot\left(\frac{f}{f_0}\right)^2=F_{DC}+\left(F_{DC}-1+NEF\right)\cdot\left(\frac{f}{f_0}\right)^2 \quad (4.29)$$

where the "dc" noise factor F_{DC} is the same as in table 4.9. Equation (4.29) shows that for values of $F_{DC}<1+NEF$, $F(f)-F_{DC}$ increases with the frequency mainly because output noise cancellation degrades (i.e. EF_{MD} increases). This highlights the importance of designing for a minimum C_1 (i.e. maximise f_0). To this purpose, the following guidelines are helpful:

- Minimize ESD and input bond-pad capacitance.
- Decrease the g_m/I_D of the MOST. However, this results in larger power dissipation.
- Use cascoding (within amplifier "Av") in order to reduce the Miller's effect.
- Use a more advanced CMOS process with a larger f_T. Since MOST capacitances scale at each new technology generation, the frequency limitations will move forward.

In order to validate the above analysis, noise simulations of the amplifier in figure 4.38a have been performed. Figure 4.39 shows the simulated matching-device noise trans-impedance versus frequency for some values of the input capacitor C_{EXTRA}. The latter is intentionally added to the circuit to account for any parasitic capacitance other than that of the matching stage, so in this case $C_1=C_{EXTRA}+C_{1,MD}$ holds and $C_{1,MD}$ is the input capacitor of the matching stage. The matching stage provides $Z_{IN}=R_S=50\Omega$ and a *noiseless* amplifier "Av" is chosen with a gain $-Av=1+R/R_S=1+300/50=7$, which ensures the dc-cancellation. The noise trans-impedance increases with the frequency even if C_{EXTRA} is zero due to $C_{1,MD}$. However, this effect is stronger as C_{EXTRA} is increased. This corresponds to a fast degradation of the noise cancellation. Figure 4.40 shows the simulated NF versus frequency for some values of C_{EXTRA}. As expected, NF increases rapidly with C_{EXTRA} and frequency because the noise trans-impedance degrades.

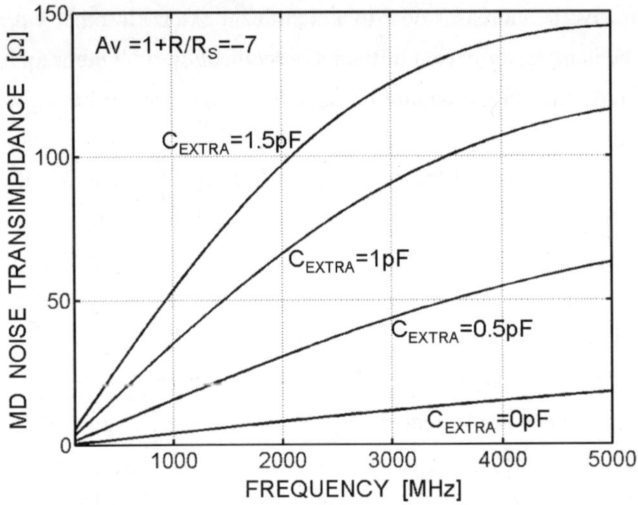

Figure 4.39: Simulated matching-device noise trans-impedance versus frequency for different values of C_{EXTRA}.

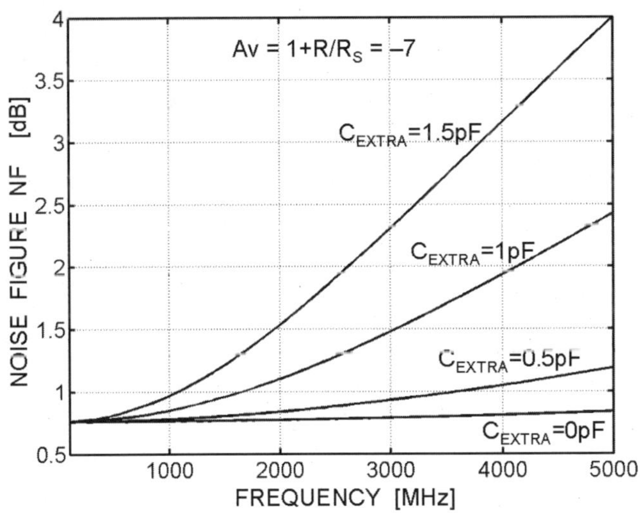

Figure 4.40: Simulated NF versus frequency for different values of C_{EXTRA}.

The above simulations clearly show a major-role played by the LNA input capacitance in determining the high-frequency degradation of the noise cancellation and so the rise of the noise factor with the frequency. However, if one tries to use equation (4.29) to fit the simulation in figure 4.40 (taking care to remove the noise contribution of "Av"), one finds that it overestimates the increases of F to a significant extent. In other words, even if the input capacitance is the main reason of the increase of EF_{MD} with the frequency, the effect of other parasitic capacitances loading the matching stage cannot be neglected if a more precise quantitative prediction is desired.

In order to investigate the actual frequency dependence of EF_{MD} the model in figure 4.38b is instead used. In this case, capacitors C_2 and C_3 are introduced to take into account the parasitic associated with matching device (C_2 accounts for the loading of the device(s) performing the signal addition). For instance, C_3 represents the gate-drain capacitance of the matching device (i.e. mainly the overlap contribution as M1 is biased in its saturation region). This relatively simple model is quite realistic because:

- Amplifier "Av" can be implemented without internal nodes (e.g.: see figure 4.16f).
- The load impedance does not affect does not affect the F of the LNA standalone.

From inspection of figure 4.38b, the noise factor can be written as:

$$F(s) = 1 + EF_M(s) + EF_R(s)$$
$$EF_M(s) = NEF \cdot g_{m1} R_S \cdot \frac{\left| Av + 1 + \frac{R}{R_S} + s \cdot R \cdot [C_1 + C_3(1 + Av)] \right|^2}{\left| Av + 1 - Rg_{m1} + s \cdot R \cdot [C_3 + Av(C_3 + C_2)] \right|^2} \tag{4.30}$$
$$EF_R(s) = \frac{R_S}{R} \cdot \frac{\left| s \cdot R \cdot [C_2 Av - C_1] - \frac{R}{R_S} - g_{m1} R \right|^2}{\left| Av + 1 - Rg_{m1} + s \cdot R \cdot [C_3 + Av(C_3 + C_2)] \right|^2}$$

for $R = -(1+Av) \cdot R_S$, $g_{m1} R_S = 1$ and $s = j\omega$ equation (4.30) is:

$$EF_{MD}(\omega) = NEF \cdot \frac{\frac{\omega^2 R_S^2}{4} [C_1 + C_3(1 + Av)]^2}{1 + \frac{\omega^2 R_S^2}{4} [C_3 + Av(C_3 + C_2)]^2}$$
$$EF_R(\omega) = -\frac{1}{1 + Av} \cdot \frac{1 + \frac{\omega^2 R_S^2}{4} [C_1 - C_2 Av]^2}{1 + \frac{\omega^2 R_S^2}{4} [C_3 + Av(C_3 + C_2)]^2} \tag{4.31}$$

Equations (4.31) shows that the frequency behaviour of $EF_{MD}(\omega)$ is different from that predicted when considering only C_1. In this case, capacitors C_2 and C_3 cause $EF_{MD}(\omega)$ to rise with a lower rate than that predicted by equation (4.28). This behaviour is originated by: (a) a *negative* input Miller capacitance $C_3(1+Av)<0$, which subtract to C_1 and (b) the term at the dominator of $EF_{MD}(\omega)$, which is originated by the fact that C_2 and C_3 create a frequency zero in the signal gain. A *negative* Miller input capacitance in the noise transfer of the matching device may appear strange. However, its existence can be easily explained. From the Miller theorem, the Miller capacitance from the input node of the LNA in figure 4.38a to ground is $C_M=C_3(1-A_{YX})$, where A_{YX} is the ratio between the voltage at node Y and the voltage at node X. Now, when the signal transfer is considered, the gain $A_{YX}=1-g_{m1}R$ is negative (for $g_{m1}R>1$) so C_M is *positive*. On the other hand, when the transfer of the noise of the matching device is considered, the situation is different. The noise voltage at node X and Y have the same sign (i.e. V_X and $V_Y=V_X(1+R/R_S)$), so the *noise* gain $A_{YX}=1+R/R_S$ is positive and larger than 1. This leads to a *negative* Miller capacitance $C_M=-C_3R/R_S$ ($=C_3(1+Av)$ for $Av=-1-R/R_S$) as far as the noise transfer of the matching device is concerned. On the other hand, C_2 increases $EF_R(f)$ at high frequencies because it leads to a *positive* Miller contribution at the numerator.

A good approximation of equations (4.31), is obtained by neglecting the term at the denominator of both $EF_{MD}(f)$ and $EF_R(f)$. In other words, the simple model of figure 4.38a can be used if one replaces C_1 with C_1-C_3R/R_S($=C_1+C_3(1+Av)$ for $Av=-1-R/R_S$) for $EF_{MD}(f)$ and C_1 with C_1-AvC_2 for $EF_R(f)$.

Figure 4.41 shows the simulated (SIM) and hand-calculated (HC) EF_{MD} of the amplifier in figure 4.38b versus frequency for some values of C_{EXTRA}. The hand calculated EF_{MD} using equation (4.28), HC1, significantly overestimates the simulation results especially for larger C_{EXTRA}. On the other hand, when C_2 and C_3 are taken into account, simulations and hand-calculation (HC123) come much closer. The same trend is observed in figure 4.42 where the noise figure NF is plotted versus frequency for some values of C_{EXTRA}. Despite the improvement, some discrepancy between simulations and hand-calculations does exist. This is due to simplification in the analysis as non quasi-static effects were neglected.

The design recommendation given to minimize $EF_{MD}(\omega)$ according to equation (4.29) are still valid when C_2 and C_3 are considered because:

- In practice, $C_1=C_{1,MD}+C_{EXTRA}$ is typically much larger than $|C_3(1+Av)|$.
- Minimizing C_1 automatically reduces also C_2 and C_3.

If after following these guidelines the high-frequency degradation of the LNA noise figure is still unacceptable, other measures such as frequency compensation must be considered. We will come back to this later.

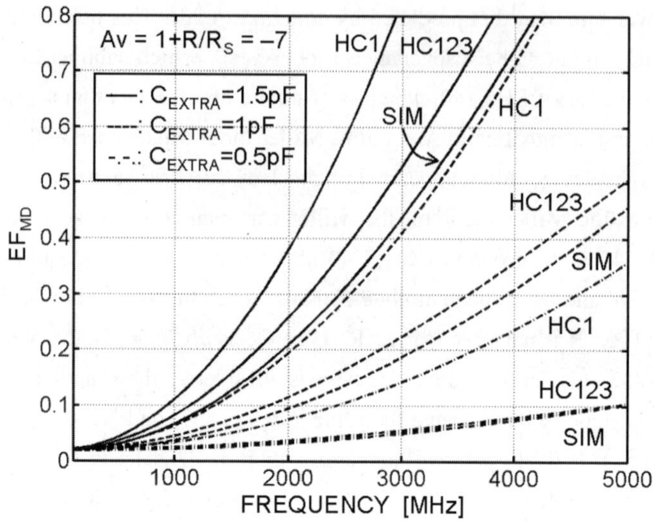

Figure 4.41: Simulated (SIM) and hand-calculated (HC) EF_{MD} versus frequency for different values of C_{EXTRA} (HC1 refers to the case when only C_1 is present and HC123 refers to case where C_1, C_2 and C_3 are present). NEF=1.37.

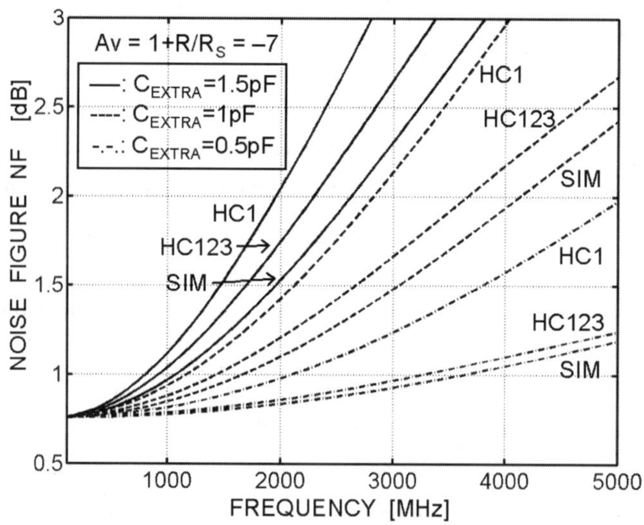

Figure 4.42: Simulated (SIM) and hand-calculated (HC) EF_{MD} versus frequency for different values of C_{EXTRA} (HC1 refers to the case when only C_1 is present and HC123 refers to case where C_1, C_2 and C_3 are present). NEF=1.37

Finally, since the input capacitance degrades the cancellation, the simultaneous noise and power matching will be affected too. This is shown in figure 4.43 where NF and NF_{MIN} of the amplifier in figure 4.38b (with R_{IN}=50Ω, –Av=1+300/50=7 and C_{EXTRA}=1pF) was simulated versus R_S for different frequencies. NF is equal at NF_{MIN} at $R_{S,OPT}$=50Ω only at 100MHz. However, up to 2 GHz the difference between NF and NF_{MIN} is relatively small.

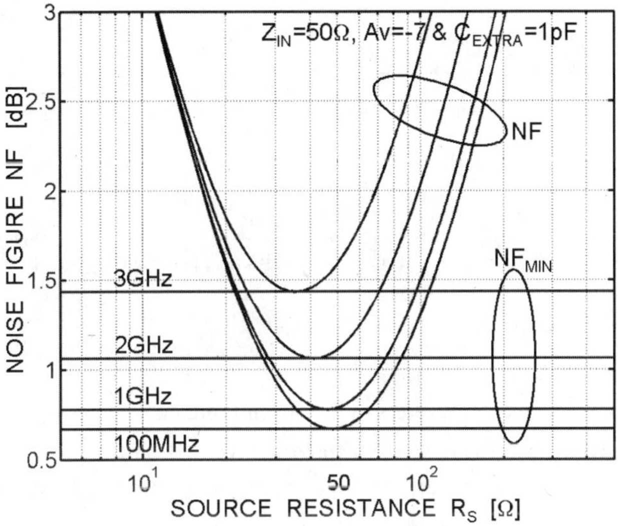

Figure 4.43: Simulated NF and NF_{MIN} of the amplifier in figure 4.38b (R_{IN}=50Ω and –Av=1+300/50=7) versus R_S at different frequencies for C_{EXTRA}=1pF.

4.9.2 Distortion

The parasitic capacitor C_1 at the input of a noise cancelling LNA degrades the distortion cancelling as well. To see it, it is sufficient to examine the 2nd and 3rd order IM distortion IMD2 and IMD3 of the amplifier in figure 4.38a for the simple case of C_2=C_3=0 and Av,2=Av,3=0 (i.e. linear amplifier "Av"). In this case, IMD2 and IMD3 can be written as:

$$IMD2_{\omega_{1\pm2}}=\frac{g_2R_SV_S}{(1+g_{m1}R_S)^2}\cdot\frac{\left|Av+1+\frac{R}{R_S}+j\omega_{1\pm2}C_1R\right|}{\left|1-g_{m1}R+Av\right|\cdot\left|1+\frac{j\omega_{1\pm2}C_1R_S}{1+g_{m1}R_S}\right|}\cdot\frac{\left|1+\frac{j\omega_{1(2)}C_1R_S}{1+g_{m1}R_S}\right|}{\left|1+\frac{j\omega_1C_1R_S}{1+g_{m1}R_S}\right|\cdot\left|1+\frac{j\omega_2C_1R_S}{1+g_{m1}R_S}\right|}$$

$$IMD3_{\omega_{2\cdot1-2}}=\frac{V_S^2}{(1+g_{m1}R_S)^3}\cdot\frac{\left|Av+1+\frac{R}{R_S}+j\omega_{2\cdot1-2}C_1R\right|}{\left|1-g_{m1}R+Av\right|\cdot\left|1+\frac{j\omega_{2\cdot1-2}C_1R_S}{1+g_{m1}R_S}\right|}\cdot\frac{\left|1+\frac{j\omega_{1(2)}C_1R_S}{1+g_{m1}R_S}\right|}{\left|1+\frac{j\omega_1C_1R_S}{1+g_{m1}R_S}\right|^2\cdot\left|1+\frac{j\omega_2C_1R_S}{1+g_{m1}R_S}\right|}\cdot$$
$$\cdot\left|-g_3R_S\frac{3}{4}+\frac{g_2^2R_S^2}{1+g_{m1}R_S}\cdot\left[\frac{1}{1+\frac{j\omega_{1\pm2}C_1R_S}{1+g_{m1}R_S}}+\frac{1/2}{1+\frac{j\omega_{2\cdot1}C_1R_S}{1+g_{m1}R_S}}\right]\right| \qquad (4.32)$$

where the notation $\omega_{1\pm2}$ and $\omega_{2\cdot1-2}$ refers to distortion products located at frequencies $\omega_1\pm\omega_2$ and $2\cdot\omega_1-\omega_2$ respectively (An analogous expression for $IMD3\omega_{2\cdot2-1}$ can also be written). Assuming "dc-cancellation" $R=-R_S\cdot(Av+1)$, $g_{m1}R_S=1$ and $f_0=1/(\pi C_1R_S)$ into equation (4.32), one gets:

$$IMD2_{f_{1\pm2}}=\frac{g_2R_SV_S}{4}\cdot\frac{\left|\frac{f_{1\pm2}}{f_0}\right|\cdot\left|1+j\frac{f_{1(2)}}{f_0}\right|}{\left|1+j\frac{f_{1\pm2}}{f_0}\right|\cdot\left|1+j\frac{f_1}{f_0}\right|\cdot\left|1+j\frac{f_2}{f_0}\right|}$$

$$IMD3_{f_{2\cdot1-2}}=\frac{V_S^2}{8}\cdot\frac{\left|\frac{f_{2\cdot1-2}}{f_0}\right|\cdot\left|1+j\frac{f_{1(2)}}{f_0}\right|}{\left|1+j\frac{f_{2\cdot1-2}}{f_0}\right|\left|1+j\frac{f_1}{f_0}\right|^2\left|1+j\frac{f_2}{f_0}\right|}\cdot\left|-g_3R_S\frac{3}{4}+\frac{g_2^2R_S^2}{2}\left[\frac{1}{1+j\frac{f_{1\pm2}}{f_0}}+\frac{1/2}{1+j\frac{f_{2\cdot1}}{f_0}}\right]\right| \qquad (4.33)$$

The voltage intercept points V_{IIP2} (at $f_{1\pm2}$) and V^2_{IIP3} (at $f_{2\cdot1-2}$) are then:

$$V_{IIP2,f_{1\pm2}} = 4 \cdot V_{IIP2,MOS} \cdot \frac{\left|\frac{f_0}{f_{1\pm2}}\right| \cdot \left|1 + j\frac{f_{1\pm2}}{f_0}\right| \cdot \left|1 + j\frac{f_1}{f_0}\right| \cdot \left|1 + j\frac{f_2}{f_0}\right|}{\left|1 + j\frac{f_{1(2)}}{f_0}\right|}$$

$$V^2_{IIP3,f_{2\cdot1-2}} = 8 \cdot \frac{\left|\frac{f_0}{f_{2\cdot1-2}}\right| \cdot \left|1 + j\frac{f_1}{f_0}\right|^2 \left|1 + j\frac{f_2}{f_0}\right| \cdot \left|1 + j\frac{f_{2\cdot1-2}}{f_0}\right|}{\left|1 + j\frac{f_{1(2)}}{f_0}\right| \cdot \left|\frac{1}{V^2_{IIP3,MOS}} + \frac{1}{2 \cdot V^2_{IIP2,MOS}} \cdot \left[\frac{1}{1 + j\frac{f_{1\pm2}}{f_0}} + \frac{1/2}{1 + j\frac{f_{2\cdot1}}{f_0}}\right]\right|} \tag{4.34}$$

where $V_{IIP2,MOS}$ and $V^2_{IIP3,MOS}$ are the "dc" intercepts of the MOS. From equations (4.34), that the intercept points of the amplifier in figure 4.38a show a strong dependence on the ratio between f_0 and the frequency of the IM product, $f_{1\pm2}$ or $f_{2\cdot1\text{-}2}$. Specifically, distortion products located well below f_0 are effectively cancelled while those near f_0 suffer from a degradation of the distortion cancellation. In addition, the ratio between f_0 and the frequency of the input tones (e.g.: f_1/f_0, f_2/f_0 and $f_{2\cdot1}/f_0$) is also important in determining the value of the intercepts. For frequencies such that $\max\{f_1, f_2, f_{1\pm2}, f_{2\cdot1\text{-}2}, f_{2\cdot1}\} << f_0$ hold, equation (4.34) further simplifies to:

$$V_{IIP2,f_{1\pm2}} = 4 \cdot V_{IIP2,MOS} \cdot \left|\frac{f_0}{f_{1\pm2}}\right|$$

$$V^2_{IIP3,f_{2\cdot1-2}} = \cdot \frac{8}{\left|\frac{1}{V^2_{IIP3,MOS}} + \frac{3}{4 \cdot V^2_{IIP2,MOS}}\right|} \cdot \left|\frac{f_0}{f_{2\cdot1-2}}\right| \tag{4.35}$$

and, in dBm

$$IIP2_{f_{1\pm2}} = 10Log_{10}\left(\frac{V^2_{IIP2,f_{1\pm2}}}{2R_S 1mW}\right) = IIP2_{MOS} + 6dB + 10Log_{10}\left|\frac{f_0}{f_{1\pm2}}\right|$$

$$IIP3_{f_{2\cdot1-2}} = 10Log_{10}\left(\frac{V^2_{IIP3,f_{2\cdot1-2}}}{2R_S 1mW}\right) = IIP3_{MOS} + 9dB - 10Log_{10}\left|1 + \frac{3}{4}\frac{V^2_{IIP3,MOS}}{V^2_{IIP2,MOS}}\right| + 10Log_{10}\left|\frac{f_0}{f_{2\cdot1-2}}\right| \quad (4.36)$$

Equation (4.36) predicts a linear in dB increase of the intercept points with the ratio $f_0/f_{1\pm2}$ and $f_0/f_{2\cdot1-2}$. In order to validate equation (4.36), the intercept points of the amplifier in figure 4.38a (with $R=-R_S\cdot(Av+1)$, $g_{m1}R_S=1$ and $Av,2=Av,3=0$) have been simulated (PSS-analysis) versus the input capacitor C_{EXTRA} (form 0 to 100pF). The input tones were located at tones $f_1=30MHz$ and $f_2=50MHz$ (each $p_{in}=-30dBm$). Figure 4.44 shows the plot of IIP2 and IIP3 versus the ratio f_0/f_d where f_d is the frequency of the related distortion product (i.e. $f_d=f_1+f_2$ for IIP2 and $f_d=2f_2-f_1$ for IIP3) and f_0 is calculated as $1/(\pi\cdot R_S\cdot(C_{EXTRA}+C_{1,MD}))$.

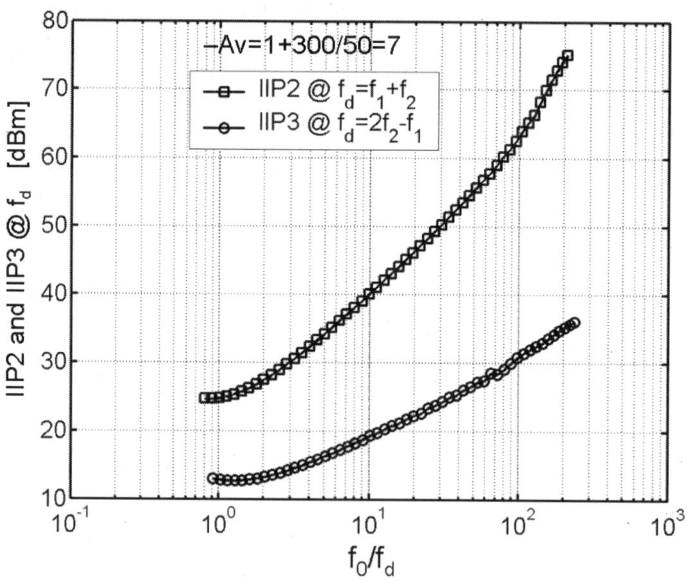

Figure 4.44: Simulated IIP2 and IIP3 versus f_0/f_d.

As can be seen from the plots, the IIPs increase linearly in dB with the ratio f_0/f_d.

4.9.3 Frequency compensation

In the previous sections, we have seen that the parasitic input capacitance of a noise cancelling LNA can cause a severe degradation of both noise and distortion cancellation at high frequency. Furthermore, some simple design guidelines were given in order to reduce this effect. When these measures do not render the desired improvement, one can apply some frequency compensation methods. In this paragraph, the focus is on noise aspects and we will tacitly assume that the compensation methods improve the distortion as well.

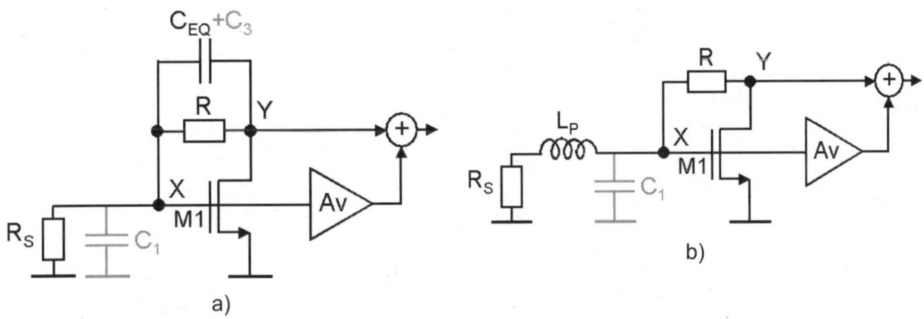

Figure 4.45: Methods to improve noise cancelling: noise equalization (a) and peaking coil (b).

We have seen that noise cancelling degrades at high frequency because of the frequency-dependent effective source impedance $Z_S(s)=R_S/(1+s\cdot R_S\cdot C_{IN})$ loads the two feed-forward noise transfer in a different fashion. Therefore, one way to improve noise cancelling is to neutralize the *different* frequency-dependence between these two paths. To do so, one can exploit the fact that, as explained in section 4.9.1, a capacitance C_{EQ} connected between nodes X and Y in figure 4.45a leads to a *negative* Miller capacitance for the noise transfer of the matching device, which counteracts the effect of C_1. From another perspective, the positive effect of C_{EQ} on noise cancelling is also evident when looking at the expressions of the two feed-forward noise transfer functions of the matching-device noise currant as:

$$
\begin{aligned}
H_{n,1,1}(s) &= \alpha(Z_S(s), g_{m1}, C_3)\cdot\left[\frac{R_S}{1+s\cdot R_S\cdot C_1}Av\right] \\
H_{n,1,2}(s) &= \alpha(Z_S(s), g_{m1}, C_3)\cdot\left[\frac{R_S}{1+s\cdot R_S\cdot C_1}+\frac{R}{1+s\cdot R\cdot\left(C_{EQ}+C_3\right)}\right]
\end{aligned}
\qquad (4.37)
$$

Upon dc-noise cancellation $R=-R_S\cdot(Av+1)$, one can write:

$$\begin{aligned} H_{n,1,1}(s) &= \alpha(Z_S(s), g_{m1}, C_3)\cdot\left[\frac{R_S}{1+s\cdot R_S\cdot C_1}Av\right] \\ H_{n,1,2}(s) &= \alpha(Z_S(s), g_{m1}, C_3)\cdot\left[\frac{R_S}{1+s\cdot R_S\cdot C_1}-\frac{R_S(1+Av)}{1-s\cdot R_S(1+Av)(C_{EQ}+C_3)}\right] \end{aligned} \tag{4.38}$$

Clearly, when the following equation is fulfilled:

$$C_1 = -(1+Av)\cdot(C_{EQ}+C_3) \;\Leftrightarrow\; C_{EQ} \equiv -\left[C_3+\frac{C_1}{1+Av}\right] > 0 \tag{4.39}$$

Equations (4.38) can be finally written as:

$$\begin{aligned} H_{n,1,1}(s) &= \alpha(Z_S(s), g_{m1}, C_3)\cdot\left[\frac{R_S}{1+s\cdot R_S\cdot C_1}Av\right] \\ H_{n,1,2}(s) &= \alpha(Z_S(s), g_{m1}, C_3)\cdot\left[\frac{R_S}{1+s\cdot R_S\cdot C_1}-\frac{R_S(1+Av)}{1+s\cdot R_S\cdot C_1}\right] = \cdots\left[-\frac{R_S Av}{1+s\cdot R_S\cdot C_1}\right] \end{aligned} \tag{4.40}$$

Equation (4.40) shows that the two feed-forward noise transfers are identical and so the cancellation can be achieved at *all frequencies even though $Z_S(s)$ remains frequency dependent.* Therefore, rearranging equation (4.31) for the amplifier of figure 4.45a as:

$$\begin{aligned} EF_{MD,EQ}(\omega) &= NEF\cdot\frac{\frac{\omega^2 R_S^2}{4}\left[C_1+(C_3+C_{EQ})(1+Av)\right]^2}{1+\frac{\omega^2 R_S^2}{4}\left[C_3+C_{EQ}+Av(C_3+C_{EQ}+C_2)\right]^2} \\ EF_{R,EQ}(\omega) &= -\frac{1}{1+Av}\cdot\frac{1+\frac{\omega^2 R_S^2}{4}\left[C_1-C_2 Av\right]^2}{1+\frac{\omega^2 R_S^2}{4}\left[C_3+C_{EQ}+Av(C_3+C_{EQ}+C_2)\right]^2} \end{aligned} \tag{4.41}$$

the numerator of $EF_{MD,EQ}$ is zero when equation (4.39) is met. Moreover, C_{EQ} improves also the frequency response of $EF_R(\omega)$ because its denominator becomes larger.

Noise equalization via C_{EQ} is very effective in reducing the high-frequency increase of EF_{MD}. It appears to be a very attractive solution. Unfortunately, this technique suffers from a drawback that can severely affect its effectiveness: capacitor C_{EQ} reduces the voltage gain of the LNA while leaving unaltered the output noise due to amplifier "Av". In other words, at high frequency, C_{EQ} increases of EF_{Av}. This degradation of EF_{Av} can be so severe that the overall LNA NF can actually be larger than for C_{EQ}=0. This can disqualify the use of noise equalization.

A more effective alternative is the use of broadband peaking-coils (i.e., shunt-peaking) as for instance shown in figure 4.45b. Here a coil L_P is connected in series to the source resistance R_S. The aim of L_P is to mitigate the frequency-dependence of the effective source impedance such that the influence parasitic input capacitor C_1 is noticeable at higher frequencies. This effect can be seen by looking at the effective source impedance for the amplifier of figure 4.45b, $Z_{S,P}(s)$:

$$Z_{S,P}(s) = \frac{s \cdot L_P + R_S}{1 + s \cdot R_S \cdot C_1 + s^2 \cdot L_P \cdot C_1} \tag{4.42}$$

For $s=j\cdot\omega$ and $\omega^2 \cdot L_P \cdot C_1 << 1$, the frequency-compensation of $Z_{S,P}(s)$ arises from the zero $\omega_z = -R_S/L_P$ that counteracts the pole $\omega_P = -1/(R_S \cdot C_1)$. However, as the frequency increases further, $\omega^2 \cdot L_P \cdot C_1 << 1$ does not hold and the frequency-dependence of $Z_{S,P}(s)$ can be worst than that of $Z_S(s)$. In other words, the peaking coil L_P improves the frequency response of EF_{MD} up to a certain frequency. Above this frequency, EF_{MD} increases even faster than for the case L_P=0 because an inductive overcompensation takes place. This is also evident from the expression of EF_{MD} for the amplifier in figure 4.45b:

$$EF_{MD,L_P}(s) = NEFg_{m1}R_S \cdot \frac{\left| Av + 1 + \frac{R}{R_S} + s^2 L_P \frac{R}{R_S}[C_1 + C_3(1+Av)] + s\left\{\frac{L_P}{R_S}(1+Av) + R[C_1 + C_3(1+Av)]\right\}\right|^2}{\left|Av + 1 - g_{m1}R + s \cdot R[C_3 + Av(C_2 + C_3)]\right|^2} \tag{4.43}$$

Substituting $R=-(1+Av)\cdot R_S$, $g_{m1}R_S=1$ and $s=j\omega$ into equation (4.43) one gets:

$$EF_{MD,L_P}(\omega) = NEF \cdot \frac{\frac{\omega^4 L_P^2}{4}[C_1 + C_3(1+Av)]^2 + \frac{\omega^2}{4}\left\{\frac{L_P}{R_S} - R_S[C_1 + C_3(1+Av)]\right\}^2}{1 + \frac{\omega^2 R_S^2}{4}[C_3 + Av(C_2 + C_3)]^2} \tag{4.44}$$

From equation (4.44), by choosing the (sub-optimal) value of L_P equal to:

$$L_P = R_S^2\left[C_1 + C_3(1+Av)\right] \tag{4.45}$$

$EF_{MD,LP}$ can be rewritten as:

$$EF_{MD,L_P}(\omega) = NEF \cdot \frac{\frac{\omega^4 R_S^4}{4}\left[C_1 + C_3(1+Av)\right]^4}{1+\frac{\omega^2 R_S^2}{4}\left[C_3 + Av(C_2+C_3)\right]^2} \tag{4.46}$$

Comparing equation (4.46) with (4.31), $EF_{MD,LP}$ is lower than EF_{MD} as far as ω is below:

$$\frac{1}{R_S\left[C_1 + C_3(1+Av)\right]} \tag{4.46}$$

Despite this limitation, up to a certain frequency, the use of peaking coil L_p renders a better noise figure response for the *overall* amplifier. This is because L_P improves (for certain frequencies) the frequency-response of the EF of *all* devices in figure 4.45b. Moreover, the frequency response of the gain and the matching are improved too.

The choice of the inductor L_P according to (4.45) is arbitrary. A better one is found searching for the value of L_P that minimizes $EF_{MD,LP}(\omega)$. In this case, one founds the following expression for L_P:

$$\frac{2}{1+\omega^2 R_S^2\left[C_3 + Av(C_2+C_3)\right]^2} R_S^2\left[C_1 + C_3(1+Av)\right] \tag{4.47}$$

The value in (4.47) is frequency-dependent and therefore it is valuable when the spot noise factor is of interest.

A better alternative to shunt peaking is to use higher-order broadband peaking-networks such as the bridged T-coil structure in figure 4.46. It consists of two inductors L_1 and L_2 with a coupling coefficient k, a bridge capacitor C_B, a load capacitor C_L at the junction point between L_1 and L_2.

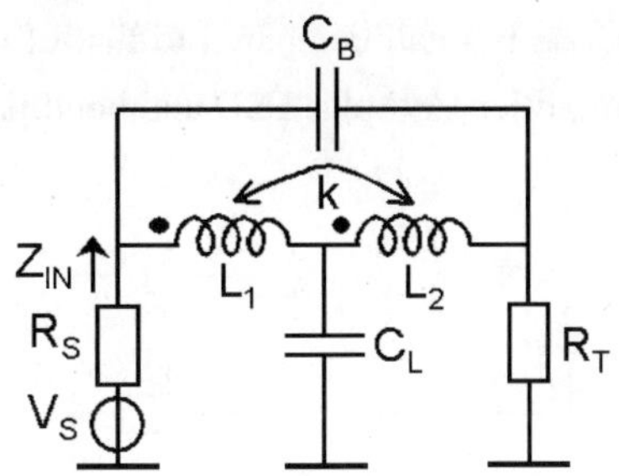

Figure 4.46: Bridged T-coil network.

This network has two important characteristics:

- Its input impedance Z_{IN} is equal to the terminating resistance R_T at *all the frequencies* and regardless the value of C_L when L_1, L_2, C_B and k are chosen according to [18]:

$$
\begin{aligned}
L_1 &= L_2 = \frac{C_L R_T^2}{4} \cdot \left(1 + \frac{1}{4\xi^2}\right) \\
C_B &= \frac{C_L}{16\xi^2} \\
k &= \frac{4\zeta^2 - 1}{4\xi^2 + 1}
\end{aligned}
\qquad (4.48)
$$

  where ξ is the damping of the poles of the network. Intuitively, this property can be recognized observing that at low frequencies the inductors are short circuits and the capacitors are open, so $Z_{IN}=R_T$. At high frequencies, the inductors are open circuits and the bridge capacitor is a short circuit, so again $Z_{IN}=R_T$.

- Its transfer bandwidth can be significantly larger than that of the shunt peaking coil. For a maximally flat group-delay response (i.e. $\xi=\sqrt{3}/2$, $L_1=L_2=C_L R_T^2/3$, $C_B=C_L/12$ and k=0.5), for instance, the bandwidth enhancement (i.e. the ratio between bandwidth of the bridged T-coil and the source bandwidth $1/R_S C_L$) of the bridged T-coil is $2\sqrt{2}=2.72$ while it is only 1.6 using shunt peaking [19].

The above properties of the bridged T-coil network can be exploited in order to improve the noise cancellation over a broader range of high frequencies compared to shunt peaking. Off course, as well as for shunt peaking, the use of the bridged T-coil network improves the frequency response of all devices EF (gain and matching too). Figure 4.47 shows a noise cancellation LNA exploiting the bridged T-coil network (with $R_T=1/g_{m1}=R_S$). It is assumed that the effect of $C_{1,MD}$ is small compared to that of C_L, which accounts for the large capacitive loading of amplifier 'Av' plus ESD and bond-pad capacitances.

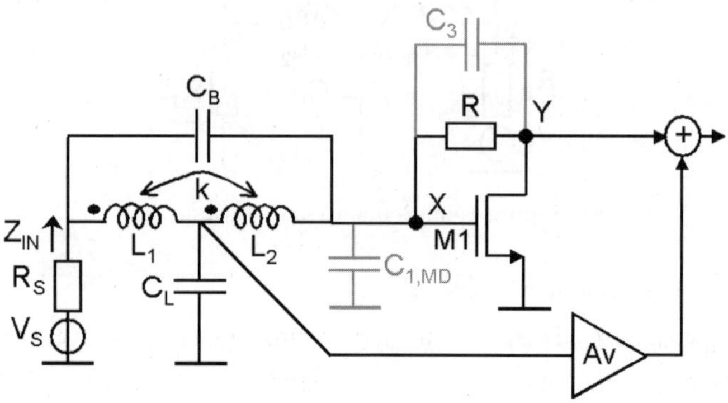

Figure 4.47: Frequency compensated noise-cancelling LNA exploiting the bridged T-coil network in figure 4.46.

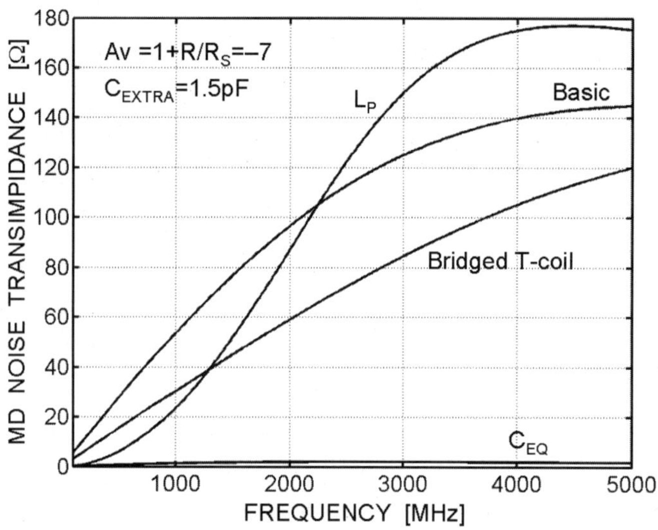

Figure 4.48: Simulated matching device (MD) noise trans-impedance versus frequency for different frequency-compensation techniques.

In order to verify the above analysis noise simulations were performed for the amplifiers in figure 4.45 and figure 4.47 designed for 'dc noise-cancellation', $Av=-1-R/R_S=-7$. To this purpose, ideal inductors were used. Moreover, an extra input capacitor $C_{EXTRA}=1.5pF$ has been added ($C_L=C_{EXTRA}$ in case of the amplifier in figure 4.47). The following values for the frequency compensation devices have been used: $C_{EQ}=226fF$, $L_P=3.4nH$ (equation

4.45), $L_1=L_2=1.24nH$, $k=0.5$ and $C_B=125fF$ (i.e. $\xi=\sqrt{3}/2$ for maximally flat group-delay response). Figure 4.48 shows the plot of the noise transfer of the matching device versus frequency for the different frequency-compensation techniques. The uncompensated response (labelled as 'basic') is also shown for sake of comparison. As expected, equalization renders a noise transfer function that is almost zero. In this respect, both the peaking networks are less effective. For the simple shunt-peaking, for instance, due to the inductive overcompensation the noise transfer function at high frequencies is actually larger than in the case without compensation.

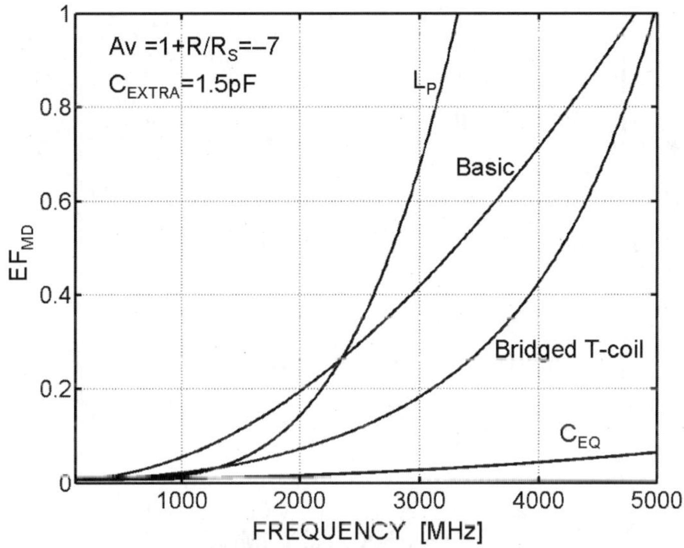

Figure 4.49: Simulated EF_{MD} versus frequency for different frequency-compensation techniques

Figure 4.49 shows the plot of EF_{MD} versus frequency for the different frequency-compensation techniques. According to the results in figure 4.48, equalization renders the lowest EF_{MD} (the residual contribution is gate induced noise), followed by the bridged T-coil and shunt-peaking. Again, above a certain frequency, simple shunt peaking has a worst EF_{MD} compared to the case without compensation.

The overall noise figure of the amplifier in figure 4.45 and figure 4.47 is plotted in figure 4.50. In order to show the effect that the different frequency compensation techniques have on all the devices EF, the noise of the local feedback resistor R and amplifier 'Av' is also considered. For amplifier 'Av', a resistor in series to the input is used to represent its equivalent input noise resistance ($R_{EQ,n,Av}$=20Ω in this case). As expected, as the noise of amplifier 'Av' is considered the effectiveness of the different frequency-compensation techniques radically changes. Noise equalization actually degrades the overall noise figure because the degradation of EF_{Av} is larger than the improvement achieved for EF_{MD}. In other words, equalization improves EF_{MD} at the expense of a larger deterioration of EF_{Av}. On the other hand, the peaking techniques remain effective also when the noise of 'Av' is present because all the devices EF_{MD} improve. Furthermore, the bridged T-coil renders the best results.

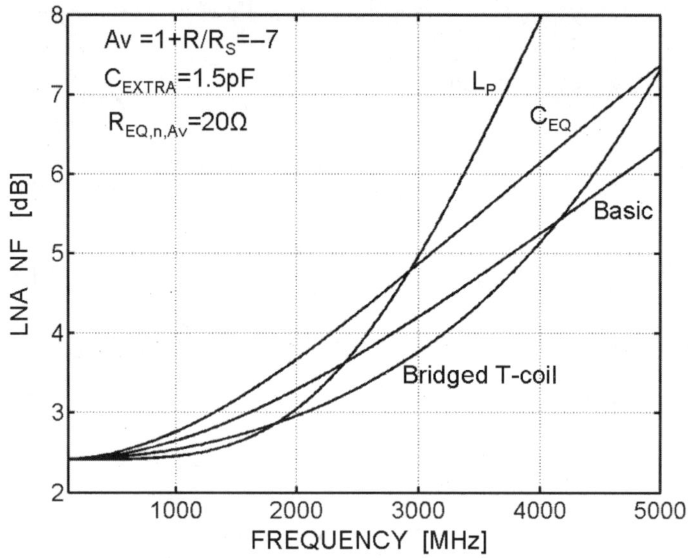

Figure 4.50: Simulated NF versus frequency for different frequency-compensation techniques.

4.10 Summary and Conclusions

Limitations to the noise performance of commonly used wide-band CMOS low-noise amplifiers have been reviewed. It was shown that:

- Elementary wide-band LNAs (figure 4.2) suffer from a fundamental trade-off between F and the impedance matching requirement $Z_{IN}=R_S$. This occurs because low values of F require the input devices to have their $g_{m,i}>>1/R_S$ and $R_i>>R_S$. On the other hand, source impedance matching demands a fixed $g_{m,i}=1/R_S$ and $R_i=R_S$. In this case, the matching device contributes to F as much as the source R_S, so F is always larger than 1+NEF>2. A somewhat lower F is obtained for a balanced common gate amplifier stage using capacitive input cross coupling (figure 4.3). Still, its noise factor upon source impedance matching is larger than 1+NEF/2 (typically between 1.5 and 2). For high sensitivity applications this value is just not enough low.

- Introducing a controlled source impedance mismatch is not an effective solution to lower NF well below 3dB.
- Practical wide-band LNAs exploiting lossy negative feedback can break the trade-off between F and $Z_{IN}=R_S$ without degrading the quality of the source match. This is done exploiting an active impedance transformation at the LNA input (e.g.: figure 4.6a), which allows for source impedance matching while generating less *equivalent* input noise than that generated by the source. In this way, noise figure values well below 3dB are possible at the price of power dissipation. Despite its noise performance, the use of negative feedback has some drawbacks too:
  - ❖ Wide-band LNAs with two or more stages in the feedback loop are prone to instability. This is especially true at high frequency where the phase shit introduced by the poles is larger.
  - ❖ The input impedance Z_{IN} depends on all the circuit parameters (e.g: $g_{m,Av}$, $R_{o,Av}$ and $G_{m,I}$ for figure 4.6a), thus: (a) Z_{IN} is sensitive to device parameter variations and (b) Since Z_{IN} and A_{VF} are directly coupled, variable gain at $Z_{IN}=R_S$ requires a more complex circuitry.
  - ❖ Furthermore, good linearity is subordinated to the availability of a sufficiently large loop gain. The latter is typically scarce at RF frequency OR it may lead to conflicting requirements. For instance, the linearity of the commonly used feedback LNA in figure 4.6a can be as poor as that of its loop amplifier Av. This is because, regardless the gain of the loop amplifier Av, the loop-gain for $Z_{IN}=R_S$ is always < 1. When amplifier Av can consist of two or more cascaded stages with most of the gain in the first one (i.e. for best noise performance), the overall linearity can be poor [4]. In fact, for this topology another trade-off exist between $Z_{IN}=R_S$ and distortion performance.

A novel wide-band noise cancellation technique was developed in order to overcome most of the above limitations. Exploiting this technique the trade-off between F and $Z_{IN}=R_S$ is broken because the output noise contribution from the impedance matching device is cancelled without degrading the signal transfer and the quality of the source match. This technique exploit the fact that two nodes of the matching stage can be found where, due to their correlation in sign, noise and signal can be distinguished from each other and so processed differently. Noise cancellation is implemented introducing a proper voltage-sensing auxiliary amplifier, which is connected to the output of the matching stage in feed-forward. Upon noise cancellation, noise figure values well below 3dB can be achieved by lowering the contribution of the "unmatched" auxiliary amplifier at the price of power dissipation. Furthermore, any undesired signal that can be modelled as a current source between the drain and source of the matching device is cancelled too (e.g: MOSFET 1/f noise, thermal noise of the distributed resistance of the gate and noise of the gate biasing). The noise cancelling technique has also been generalised to other circuit implementations. A number of alternative LNA circuit exploiting noise cancelling have been generated systematically (figure 4.16). Their noise performance has been compared based on hand calculations. It was found that the LNA in figure 4.16f has the lowest F upon $Z_{IN}=R_S$, a given power, voltage gain A_{VF} including also noise of the biasing circuitry and the supply voltage limitations. Furthermore, it is more suitable for high frequencies due to the local shunt feedback at the output of the matching stage.

Despite the differences in their noise performance, these noise-cancelling LNAs share the following peculiar properties:

- Instability risks are relaxed due to the feed-forward nature of noise cancelling.
- The cancellation relies on the absolute value of the *real* impedance of the source, R_S.
- To a first order, their input impedance Z_{IN} depends only on the matching device g_m. Therefore, Z_{IN} is less sensitive to device parameter variations (e.g.: process spread). Also, variable gain at constant $Z_{IN}=R_S$ is easier implemented.
- Noise cancelling is *robust* to device parameter variations because: (a) the solution of the noise-cancelling equation depends on a reduced-set of circuit parameters and (b) variations of the value of R_S and the gain A_{VF2} of the auxiliary (voltage-sensing) amplifier have a modest impact on the noise figure contribution of the matching device.
- Noise cancelling allows for (in-band) *simultaneous noise and power* matching without an extra wideband matching network. This occurs because, upon noise cancellation, the noise transfer functions of all devices remain equal upon short and open conditions at the input of the LNA. This is strictly true for a MOST that is regarded as a VCCS.
- Assuming a *linear* voltage-sensing auxiliary amplifier, all the output distortion components generated by the matching-device are cancelled in the same manner and upon the same conditions noise cancels at the output of the LNA. In other terms, *simultaneous matching-device noise and distortion cancellation* is achieved. When the

non-linearity of the auxiliary amplifier is introduced, different scenarios are possible depending on how its distortion components do combine at the output with that due to the matching-device. Under proper conditions, total cancellation of even *or* odd output distortion still occurs but this will not happen simultaneously with noise cancellation.

High-frequency limitations to noise cancelling have been also investigated. It was found that the effective input capacitance of a noise-cancelling LNA is the main reason for the degradation of the cancellation and the consequent increase of the noise figure. This occurs because the matching device noise current sees frequency dependent source impedance that affects the two feed-forward noise transfers in a different manner. In the same manner, this input capacitance is also responsible for the degradation of the distortion cancellation as well the simultaneous noise and power matching.

Some design guidelines has been given in order to mitigate the noise figure degradation at high frequency. One can design the LNA for minimum parasitic input capacitance using a larger g_m/I_D for the matching device, cascoding in the auxiliary amplifier to reduce the Miller effect, design ESD protection device and use an input bond-pad with low parasitic capacitance and a more advanced sub-micron CMOS process with a higher f_T. If the previous measures are ineffective, a frequency-compensation network can be used. In this respect, two options have been examined with regard to a specific LNA topology: (a) capacitive noise equalization and (b) peaking networks. The former neutralizes the frequency dependence of EF_{MD} by equalizing the effect that the complex source impedance has over the noise transfer function of the two feed-forward paths. However, the frequency-independence of EF_{MD} comes at the price of a significant degradation of the EF of the auxiliary amplifier. The net result is than an even larger LNA noise figure. Furthermore, the equalization capacitance degrades gain and matching at high frequency. More effectively, peaking coils can be used to mitigate the degradation of noise cancelling by compensating, to some extent, the frequency-dependence of the effective source impedance seen by the noise current flowing out from the matching device. The great advantage of peaking networks is that they improve the frequency dependence of *all* devices EF, which always results in an improvement of the LNA noise figure. Furthermore, the frequency response of the gain and matching versus frequency are improved too. More specifically, two cases of peaking networks have been examined: a shunt peaking coil in series to R_S and a properly connected bridged T-coil network. The former is simple to use but yields a more limited NF improvement versus frequency whereas the latter is only somewhat more complex and renders superior results.

4.11 References

[1] Q. Huang et al., "GSM Transceiver Front-end Circuits in 0.25-um CMOS", IEEE Journal of Solid State Circuits, pp.292-302, March 1999.

[2] W. Zhuo, et al., "Using Capacitive Cross-Coupling Technique in RF Low Noise Amplifiers and Down-Conversion Mixer Design", IEEE proceedings of ESSCIRC'2000, pp.116-119, Stockholm, Sweden, 19-21 September 2000.

[3] O. Watanabe, et al., "A 2GHz Down-Converter with a –3dB Bandwidth of 600MHz using LO Signal Suppressing Output Buffer", IEEE International Solid-State Circuits Conference ISSCC, slide supplement, San Francisco, California, USA, February 2001.

[4] P. Leroux, et al., "A 0.8 dB NF, ESD-protected, 9mW CMOS LNA", International Solid-State Circuits Conference, San Francisco, USA, pp.410-411, February 2001.

[5] Norton, D. E., "High Dynamic Range Feedback Amplifiers Using Loss-less Feedback", Proceedings IEEE Symposium on Circuits and Systems, pp. 438-440, 1975.

[6] Rhode, U.L. "The Design of a Wide-Band Amplifier with Large Dynamic Range and Low Noise Figure Using CAD Tools", IEEE Symposium on Microwave Theory and Techniques, pp. 47-55, April 1987.

[7] E. Nordholt "The Design of High-performance Negative Feedback Amplifiers", Elsevier Scientific Publishing, Amsterdam, 1983.

[8] A. Zolfaghari et al., "Stacked Inductors and Transformers in CMOS Technology", IEEE Journal of Solid State Circuits, vol. 36, No. 4, pp.292-302, April 2001.

[9] M. Martin, "Ferrite Transformers Minimise Losses in RF Amplifiers", Microwaves & RF Journal, pp.117-126, May 1990.

[10] C. D. Hull and R. G. Meyer et al., "Principles of Monolithic Wide-band Feedback Amplifier Design", International Journal of High Speed Electronics, vol. 3, No. 1, pp.53-93, March 1992.

[11] J. Janssens et al., "A 2.7Volt CMOS Broad Band Low Noise Amplifier", Symposium on VLSI circuits, pp. 87-88, Kyoto, June 1997.

[12] K. Bult, "Analog CMOS Square-Law Circuits", Ph.D. Thesis, University of Twente, Enschede, The Netherlands, 1988.

[13] F. Piazza, et al., "A High Linearity, Single-Ended Input Double-Balanced Mixer in 0.25μm CMOS", IEEE proceedings ESSCIRC'98, pp.60-63, The Hague, The Netherlands, 22-24 September, 1998.

[14] H. Seidel, "A Microwave Feed-Forward Experiment", Bell Systems Technical Journal, vol.50, pp.2879-2916, November 1971.

[15] Y. Sherbelis et. Al, "Microwave Feed-Forward Low Noise Amplifier Design for Cellular Base Station", Electronics Letters, vol. 37, No. 6, pp. 359-361, March 2001.

[16] H. Rothe and W. Dahlke, "Theory of Noisy Four poles", Proceedings IRE, vol.44, pp.811-818, June 1956.

[17] P. Wambacq and W. M. Sansen, "Distortion Analysis of Analog Integrated Circuits", Kluwer Academic Publisher, July 1998.

[18] L. Selmi, D.B. Estreich and B. Ricco`, "Small-signal MMIC Amplifiers with Bridged T-Coil Matching Networks", IEEE Journal of Solid State Circuits, vol. 27, No 7, pp.1093-1096, July 1992.

[19] S. Galal and B. Razavi "Broadband ESD Protection Circuits in CMOS Technology", International Solid-State Circuits Conference, San Francisco, USA, pp.182-183, February 2003.

Chapter 5
Design of a Decade Bandwidth Noise Cancelling CMOS LNA

5.1 Introduction

In chapter 4, a novel feed-forward low-noise cancelling technique has been introduced. By exploiting this technique, input-matched low-noise amplifiers with noise figures well below 3dB over a wide range of frequencies can be designed without using negative feedback. In this chapter, noise cancelling is demonstrated through the design of a wide-band LNA in standard 0.25μm digital CMOS.

5.2 Design Requirements and LNA Schematic

This paragraph deals with the choice of the LNA topology and the design requirements. Since the aim is to prove the new concept, the design will focus on the noise performance[XV] and frequency behaviour. To this purpose, we will address high-sensitivity applications with demanding requirements for the noise figure and gain of the LNA. In this respect, the following target requirements will be used:

- Wide signal bandwidth: from MHz to GHz while driving a capacitive load.
- Voltage gain: $A_{VF}=V_{OUT}/V_{IN}=10$.
- Source impedance matching: $Z_{IN}=R_S=50\Omega$, return loss larger than 8-10 dB.
- Noise Figure: well below 3dB over the signal bandwidth.

Figure 5.1 shows the schematic of the noise-cancelling LNA, which is based on the amplifier circuit in figure 4.16f. The latter is preferred to the other noise cancelling LNAs because it has lower NF and good high-frequency properties due to the (local) shunt-feedback in the matching stage. The characteristics of this LNA are the followings:

- The matching stage exploits local shunt-feedback around a CMOS inverter providing an input impedance $Z_{IN}=1/(g_{m1a}+g_{m1b})$. The inverter has a larger g_m/I_D compared to a common source MOS because the p-type MOST reuses the bias current of the n-type MOST. On the other hand, its input capacitance is typically larger. To lower sensitivity

[XV] Distortion cancelling will not be exploited because, at that time, its existence was recognised.

of gain and input impedance to variations in the supply voltage, the inverter is biased via a current mirror while a MOS capacitor C_1 grounds the source of M1b.

- The inverter output is ac-coupled to M3 via a first-order high-pass filter, C_2-R_2.
- The cascode device M2b improves reverse isolation and decreases the input capacitance by reducing the Miller effect around M2a.
- The bias current, I_{BIAS2}, allows M3 to conduct a part of the current of M2. This allows a lower supply voltage (V_{DD}=2.5V) without sacrificing NF because the LNA gain is large and enough voltage headroom is available across the current mirror output.
- The capacitance C_{PAD} of the output bond-pad is used to emulate the load capacitance.

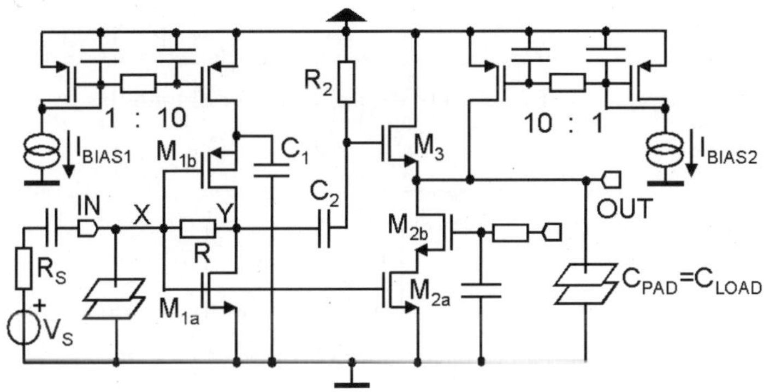

Figure 5.1: Schematic of the selected wide-band LNA circuit.

5.3 Analysis of Noise Factor and Bandwidth

In this paragraph, the noise factor and the bandwidth of the LNA is analysed in depth. The aim is to derive equations that will be used to optimise its noise performance.

In chapter 4, the LNA noise factor was analysed under the assumption that *exact* noise cancellation (i.e. $g_{m2}/g_{m3}=1+R/R_S$ for the LNA in figure 5.1) leads to the lowest possible F. In the following, the aim is to investigate if a *partial* cancellation can be beneficial to decrease the noise factor further. The analysis of F is divided in two parts concerning the in-band and high-frequency behaviour respectively. Furthermore, the MOST is no longer regarded as VCCS and only channel noise of devices in the signal path is considered as:

- The corner frequency of the output 1/f noise is well below the MHz range as:
  - The inverter 1/f output noise is cancelled.
  - The 1/f output noise due to M2a is relatively low due to its large size.
- The resistance of the gate terminal of the MOSFETs and the substrate underneath can be significantly reduced by a proper layout practice [1,2].

- Noise generated by the bias circuitry does not affect F significantly because:
  - The noise generated by I_{BIAS1} cancels to the output.
  - LNA gain is large and large voltage headroom is available to the output mirror.
- The output noise from the HPF C_2-R_2 can be made small by increasing C_2 and R_2.
- A well-designed cascode MOST contributes a little to F.

5.3.1 F for in-band frequencies

At frequencies where the effect of circuit capacitances can be neglected the noise factor of the LNA in figure 5.1 can be written as $F = 1 + EF_{CS} + EF_{MD} + EF_R$ with:

$$EF_{CS} = \frac{NEF \cdot \left[\frac{g_{m2}}{g_{m3}} + \left(\frac{g_{m2}}{g_{m3}}\right)^2\right] \cdot \left(1 + \frac{g_{m1}R_S + \frac{R_S}{R_O}}{1 + \frac{R}{R_O}}\right)^2}{g_{m2}R_S \cdot \left(\frac{1 - g_{m1}R}{1 + \frac{R}{R_O}} - \frac{g_{m2}}{g_{m3}}\right)^2}$$

$$EF_{MD} = \frac{NEF \cdot g_{m1}R_S}{\left(\frac{1 - g_{m1}R}{1 + \frac{R}{R_O}} - \frac{g_{m2}}{g_{m3}}\right)^2} \cdot \frac{\left(\frac{R}{R_S} + 1 - \frac{g_{m2}}{g_{m3}}\right)^2}{\left(1 + \frac{R}{R_O}\right)^2}$$

$$EF_R = \frac{\frac{R}{R_S} \cdot \left(\frac{1 + g_{m1}R_S + \frac{g_{m2}}{g_{m3}} \cdot \frac{R_S}{R_O}}{1 + \frac{R}{R_O}}\right)^2}{\left(\frac{1 - g_{m1}R}{1 + \frac{R}{R_O}} - \frac{g_{m2}}{g_{m3}}\right)^2} \tag{5.1}$$

EF_{CS}, EF_{MD} and EF_R are the excess noise factor of the common-source stage, the matching device and the resistor R, respectively. Furthermore, $g_{m1}=g_{m1a}+g_{m1b}$ and $R_O=1/(g_{d1a}+g_{d1b})$

are the transconductance and output resistance of the inverter. NEF is assumed constant and equal for all the MOSTs.

The gain, A_{VF}, and input impedance, Z_{IN}, for g_{mb3}>>g_{d3} and $g_{d2a}/(1+g_{m2b}/g_{d2b})$<<$g_{d3}$ are:

$$A_{VF}=\left(\frac{1-g_{m1}R}{1+\frac{R}{R_O}}-\frac{g_{m2}}{g_{m3}}\right)\cdot\frac{1}{1+n} \quad \text{and} \quad Z_{IN}=\frac{1+\frac{R}{R_O}}{g_{m1}+\frac{1}{R_O}} \tag{5.2}$$

where $n=g_{mb3}/g_{m3}$ is a constant (typically between 0 and 0.5), which accounts for the body-effect of M3. From equation (5.1)-(5.2), we can conclude that:

- Given g_{m1}, R, g_{m2} and g_{m3}, R_O increases Z_{IN} while both R_O and 'n' decreases A_{VF}.
- The noise cancelling condition, $g_{m2}/g_{m3}=1+R/R_S$, is independent on the value of R_O and the g_d and g_{mb} of M2 and M3. This can be understood, since these parameters affect equally both noise paths.
- For $Z_{IN}=R_S$ and $g_{m2}/g_{m3}=1+R/R_S$, EF_R is equal to R_S/R independently on the value of R_O (and n)[XVI].
- From previous reasoning, g_{d3} and g_{mb3} do not directly affect the value of F.

Using equations (5.1) and (5.2), F for $Z_{IN}=R_S$ can be rearranged as a function of the parameters $\{g_{m2}R_S, \varepsilon, A_{VF}, R_O \text{ and } n\}$ as follows:

$$EF_{CS}\equiv\frac{NEF}{g_{m2}R_S}\cdot\frac{\varepsilon^2+\varepsilon\cdot(6-2A_{VF}(1+n))+4-6A_{VF}(1+n)+A_{VF}^2(1+n)^2}{A_{VF}^2(1+n)^2}$$

$$EF_{MD}\equiv\frac{NEF\cdot\varepsilon^2}{A_{VF}^2(1+n)^2}\cdot\frac{1-\frac{(\varepsilon+2+A_{VF}(1+n))R_S}{2R_O}}{\left(1-\frac{(\varepsilon+A_{VF}(1+n))R_S}{2R_O}\right)^2}$$

$$EF_R\equiv-2\cdot\frac{\varepsilon+A_{VF}(1+n)}{A_{VF}^2(1+n)^2}\cdot\left(\frac{1-\frac{A_{VF}(1+n)R_S}{2R_O}}{1-\frac{(\varepsilon+A_{VF}(1+n))R_S}{2R_O}}\right)^2$$

[XVI] The noise current of R can be divided in two correlated sources, from the input to ground and the drain of M1 to ground. The latter renders no output noise, the contribution to F of the former is $I_{n,R}^2/I_{n,RS}^2$.

$$A_{VF} = -\frac{2\dfrac{R}{R_S}+\varepsilon}{1+n} \quad \text{and} \quad g_{m1}R_S = 1+\frac{R-R_S}{R_O} \tag{5.3}$$

In the above equations, the *noise cancelling error*, $\varepsilon \equiv g_{m2}/g_{m3}-1-R/R_S$, is defined in order to quantify the amount of cancellation of matching device noise at the output of the LNA. For instance, exact noise cancellation occurs for $\varepsilon=0$. In this case, equations (5.3) can be rewritten as follows:

$$\begin{aligned} EF_{CS,\varepsilon=0} &= \frac{NEF}{g_{m2}R_S}\cdot\frac{4-6A_{VF}(1+n)+A_{VF}^2(1+n)^2}{A_{VF}^2(1+n)^2} \\ EF_{MD,\varepsilon=0} &= 0 \quad \text{and} \quad EF_{R,\varepsilon=0} = \frac{-2}{A_{VF}(1+n)} = \frac{R_S}{R} \end{aligned} \tag{5.4}$$

The effect of a partial cancellation on the noise factor can now be investigated by looking the behaviour of EF_{MD}, EF_R and EF_{CS} for $\varepsilon \neq 0$. These quantities are plotted in figure 5.2 versus the pair (ε, $g_{m2}R_S$) assuming NEF=1.5, A_{VF}=-10, R_O=1KΩ and n=0.2[XVII].

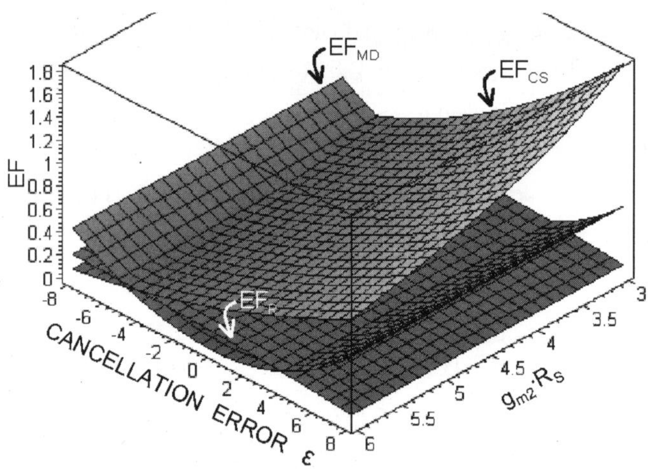

Figure 5.2: EF_{MD}, $EF_R$and EF_{CS} versus the pair (ε, $g_{m2}R_S$).

EF_{MD} increases as ε differs from zero due to imperfect noise cancellation. EF_R decreases for $\varepsilon>0$, while it increases for $\varepsilon<0$. In both cases, the variation of EF_R is rather modest. On the other hand, EF_{CS} drops substantially for $\varepsilon<0$. This is because for a given A_{VF} and g_{m2},

[XVII] R_O=1KΩ is obtained for $g_m=1/R_S$, $V_{gs}=V_{DS}=0.25+V_{T0}$ and $L=L_{min}$.

$\varepsilon<0$ leads to a smaller $1/g_{m3}=(1+\varepsilon/2-A_{VF}(1+n)/2)/g_{m2}$, thus lowering the output noise of M2-M3 (i.e. EF_{CS} drops). From previous reasoning, it can be argued that the noise factor can assume values below $F(\varepsilon=0)$ for some $\varepsilon<0$ such that the drop of $EF_{CS}(\varepsilon)$ dominates the increase of $EF_{MD}(\varepsilon)$ and $EF_R(\varepsilon)$.This optimal value of ε, ε_{OPT}, is shown in figure 5.3 where contour lines of $NF(\varepsilon, g_{m2}R_S)$ for $Z_{IN}=R_S$ are plotted assuming NEF=1.5, A_{VF}=-10, R_O=1KΩ and n=0.2. Figure 5.4 shows the same contour lines for $R_O=\infty$ and n=0.

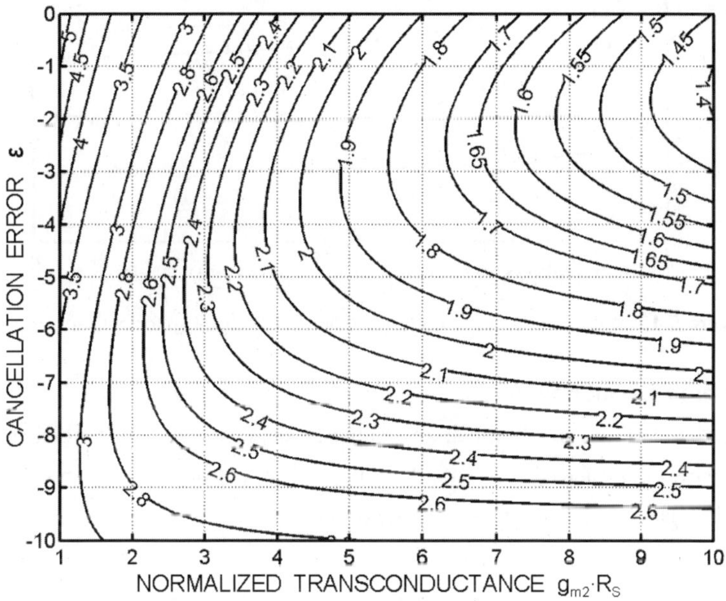

Figure 5.3: Contour lines of the LNA noise figure $NF(\varepsilon,g_{m2}R_S)$ for R_O=1KΩ and n=0.2.

From both these figures, we conclude that:

- For a given $g_{m2}R_S$, NF for $\varepsilon=\varepsilon_{OPT}$ is lower than that for $\varepsilon=0$. For instance, $g_{m2}R_S=4$ and $\varepsilon_{OPT}\approx-4$ lead to a $NF(\varepsilon=0)-NF(\varepsilon_{OPT})$ as large as 0.35dB for equal power dissipation!
- For a given NF, $\varepsilon=\varepsilon_{OPT}$ requires the lowest value of $(g_{m2}R_S)_{OPT}$, thereby reducing the power dissipation. For instance, for NF≈2.05dB, $\varepsilon=0$ requires $g_{m2}R_S\approx5$ while $\varepsilon=\varepsilon_{OPT}\approx-4$ requires only $g_{m2}R_S=(g_{m2}R_S)_{OPT}\approx4$.
- Also, $\varepsilon=\varepsilon_{OPT}$ occurs for $\partial F/\partial\varepsilon=0$, which means a lower sensitivity to device parameters variations (i.e. process-spread). However, ε_{OPT} is itself sensitive to second-order effects as can be seen from the difference between figure 5.3 and 5.4.

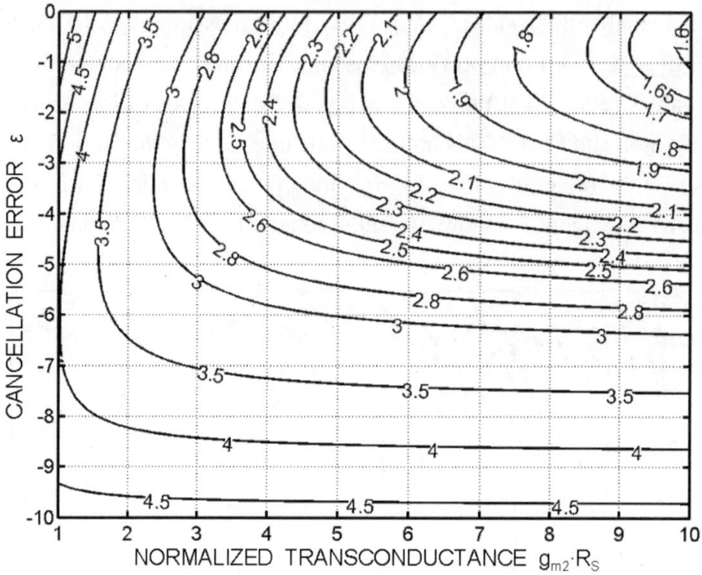

Figure 5.4: Contour lines of the LNA noise figure NF(ε,$g_{m2}R_S$) for n=0 and $R_o=\infty$.

5.3.2 F at high frequencies

As discussed in section 4.9.1, the high-frequency degradation of the noise factor $F(f) = 1+EF_{CS}(f)+EF_{MD}(f)+EF_R(f)$ is essentially described by considering the specific *effective* parasitic capacitance $C_{IN,eff}$ at the input node of the LNA as:

$$EF_{CS}(f) = EF_{CS}\cdot\left[1+\left(\frac{2\pi f\cdot C_{IN,eff,CS}R_S\left(1+\dfrac{R}{R_O}\right)}{1+g_{m1}R_S+\dfrac{R+R_S}{R_O}}\right)^2\right]$$

$$EF_{MD}(f) = EF_{MD}\cdot\left[1+\left(\frac{2\pi f\cdot C_{IN,eff,MD}R}{1+\dfrac{R}{R_S}-\dfrac{g_{m2}}{g_{m3}}}\right)^2\right]$$

$$EF_R(f) = EF_R\cdot\left[1+\left(\frac{2\pi f\cdot C_{IN,eff,R}R_S}{1+g_{m1}R_S+\dfrac{g_{m2}R_S}{g_{m3}R_O}}\right)^2\right] \quad (5.5)$$

Expressions for these effective input capacitances are:

$$C_{IN,eff,MD} \approx C_{gs,1} + C_{gs,2} + C_{gb,1} + C_{gb,2} + 2C_{gd,2} + C_{gd,1}\left(1 - \frac{g_{m2}}{g_{m3}}\right)$$

$$C_{IN,eff,R} \approx C_{gs,1} + C_{gs,2} + C_{gb,1} + C_{gb,2} + 2C_{gd,2} + C_{db,1}\frac{g_{m2}}{g_{m3}} \quad (5.6)$$

$$C_{IN,eff,CS} \approx C_{gs,1} + C_{gs,2} + C_{gb,1} + C_{gb,2} + 2C_{gd,2} + C_{gd,1}\left(1 + \frac{g_{m1}R}{1 + \frac{R}{R_0}}\right)$$

where the bond-pad and ESD capacitances were neglected. After some manipulations, equations (5.6) can be rewritten in terms of the intrinsic MOSFET $C_{gs1,n}$ and $C_{gs2,n}$ and the technology parameters [1]: $\alpha_{np}=\mu_n/\mu_p$, $\alpha_{ov,x}=C_{ov}/C_{gs,x}$ and $\alpha_{jn,x}=C_{jx}/C_{gs,x}$. Finally, using (5.3) and (5.6) , equations (5.5) can be written as:

$$EF_{CS}(f) = EF_{CS} \cdot \left[1 + \chi_{CS} \cdot \left(\frac{f}{f_T}\right)^2\right]$$

$$EF_{MD}(f) = EF_{MD} \cdot \left[1 + \chi_{MD} \cdot \left(1 + \frac{A_{VF}(1+n)}{\varepsilon}\right)^2 \cdot \left(\frac{f}{f_T}\right)^2\right] \quad (5.7)$$

$$EF_R(f) = EF_R \cdot \left[1 + \chi_R \cdot \frac{\left(\frac{f}{f_T}\right)^2}{\left(1 - \frac{A_{VF}(1+n)R_S}{2R_O}\right)^2}\right]$$

where $f_T \approx g_m/C_{gs}$ is the unity gain frequency of the input nMOST (i.e. M1a and M2a) and χ_{CS}, χ_{MD} and χ_R are a function of R_O, A_{VF}, n, ε and the parameters α_{np}, $\alpha_{ov,x}$ and $\alpha_{jn,x}$. The EF's and NF for ε=0 versus ($g_{m2}R_S$, f/f_T) are shown in figures 5.6 and 5.7. From these figures, it follows that:

- For large values of $g_{m2}R_S$, $EF_{MD}(f)$ increases faster with f/f_T than $EF_R(f)$ and $EF_{CS}(f)$ do, thereby determining a degradation of NF(f). For f/f_T close to 0.1, the increase of NF can be such that frequency compensation is required (see chapter 4). Nevertheless, for a 0.25μm CMOS process with an f_T of 40GHz, f/f_T=0.1 renders more than 4GHz.

- The previous consideration highlights the importance of minimising C_{IN} by increasing the f_T using a large V_{GS}-V_{T0} for M1 and M2. For a given g_m, a larger V_{GS}-V_{T0} increases the power consumption $P \approx P_{MOST,CS}(g_{m1}R_S+g_{m2}R_S) \approx (V_{DD}(V_{GS}-V_{T0})/R_S)(0.5+g_{m2}R_S)/2$.

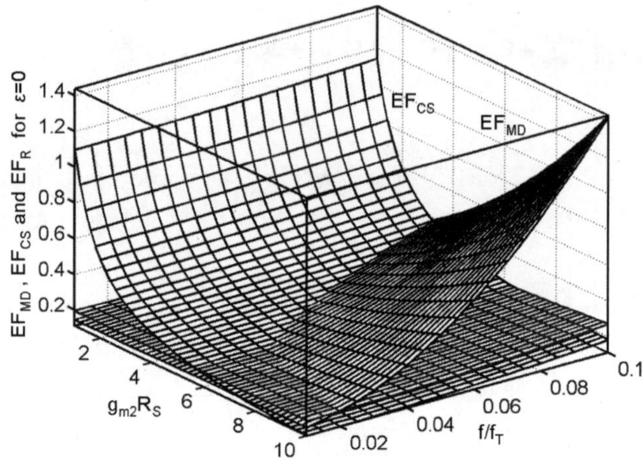

Figure 5.6: EF's at $\varepsilon=0$ versus ($g_{m2}R_S$, f/f_T) for A_{VF}=-10, n=0.2, NEF=1.5, R_O=1KΩ, L_{min}= 0.25μm, $\alpha_{ov,n}$=0.2, $\alpha_{ov,p}$=0.5 α_{jn}=0.32, α_{jp}=0.28 and α_{np}=2.5.

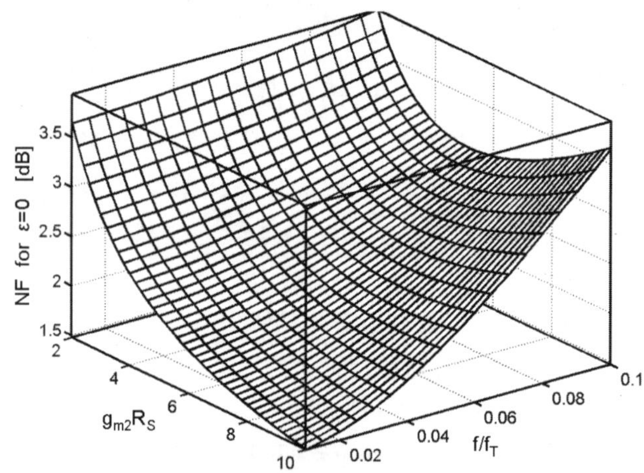

Figure 5.7: NF at $\varepsilon=0$ versus ($g_{m2}R_S$, f/f_T) for A_{VF}=-10, n=0.2, NEF=1.5, R_O=1KΩ, L_{min}=0.25μm, $\alpha_{ov,n}$=0.2, $\alpha_{ov,p}$=0.5, α_{jn}=0.32, α_{jp}=0.28 and α_{np}=2.5.

Figure 5.8 and 5.9 show the device's EF and NF versus ε for $g_{m2}R_S$=5 and f/f_T={0.01, 0.05, 0.09}. NF degrades with f/f_T more when $-\varepsilon$ increases due to imperfect cancellation. From these figures we can conclude that, for $\varepsilon=\varepsilon_{OPT}$, the in-band NF decreases compared to NF(ε=0) at the price of a larger high-frequency degradation. This degradation can be mitigated if a frequency compensation network is applied at the input (see chapter 4).

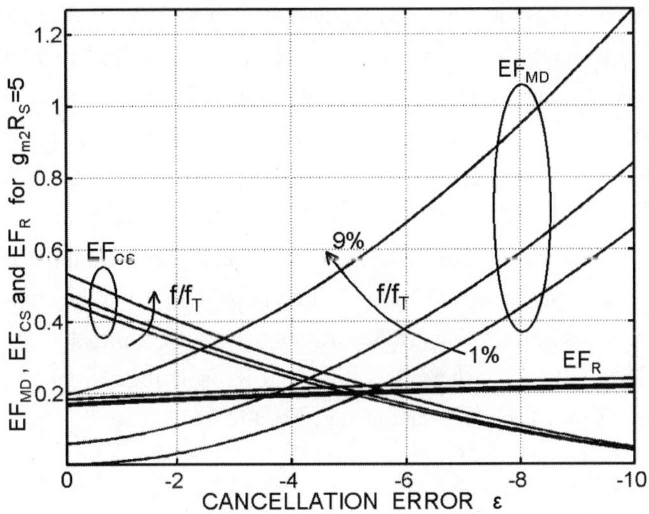

Figure 5.8: EF's versus ε=0 for $g_{m2}R_S$=5 and f/f_T= 0.01, 0.05 and 0.09.

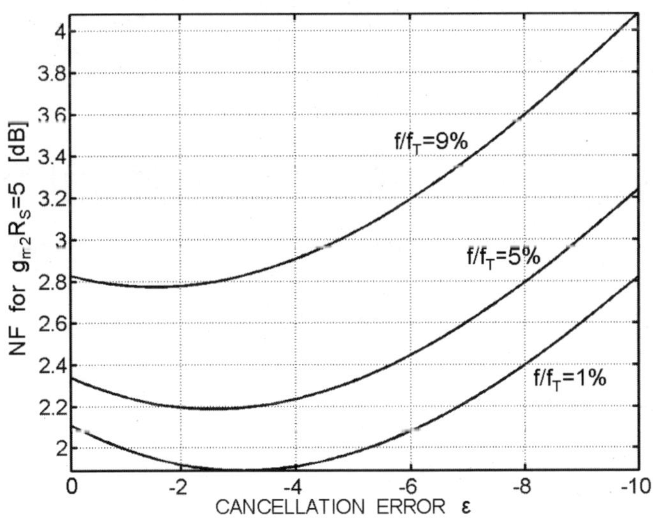

Figure 5.9: NF versus ε=0 for $g_{m2}R_S$=5 and f/f_T = 0.01, 0.05 and 0.09.

5.3.3 Bandwidth

In this paragraph, a simplified analysis of the bandwidth of the LNA in figure 5.1 is provided exploiting the method of the open time-constant [2,3]. This method provides accurate results when: (a) the input output transfer function is made by real poles and (b) one of the poles is dominant. For the LNA in figure 5.1, both the previous assumptions may not hold. For instance, a zero in the input-output transfer function occurs when the feed-forward paths contain different dominant poles. However, the application of the method to these cases typically leads to conservative estimates of the bandwidth [2,3]. Next, the method is simple and has the merit to show what capacitances limit the bandwidth. In the following analysis, it is further assumed that important capacitances are referred to ground: at the input node (C_{IN}), output node (C_{OUT}) and the output node of the matching stage (C_{CSSF}). The related poles are $p_{IN}=-1/(2\pi R_{IN}C_{IN})$, $p_{OUT}=-1/(2\pi R_{OUT}C_{OUT})$ and $p_{CSSF}=-1/(2\pi R_{CSSF}C_{CSSF})$, where R_{IN}, R_{OUT} and R_{CSSF} is the resistance seen from the respective capacitances. The –3dB bandwidth is then [2,3]:

$$f_{-3dB} \approx -\frac{1}{\frac{1}{p_{IN}}+\frac{1}{p_{OUT}}+\frac{1}{p_{CSSF}}} \tag{5.8}$$

From the inspection of figure 5.1, after some manipulations, the poles can be written as:

$$\left\{ p_{IN} = -\frac{f_T}{\chi_{IN}};\quad p_{OUT} = -\frac{f_T}{\chi_{OUT}};\quad p_{CSSF} = -\frac{f_T}{\chi_{CSSF}} \right\} \tag{5.9}$$

with

$$\chi_{IN} = \left(1-\frac{R_S}{R_O}\cdot\frac{2+A_{VF}(1+n)+\varepsilon}{2}\right)\cdot\frac{1+\alpha_{np}+\alpha_{ov,n}+\alpha_{ov,p}\alpha_{np}}{4}+\frac{g_{m2}R_S\left(1+2\alpha_{ov,n}\right)}{2}$$

$$\chi_{OUT} = \left(1+\frac{\varepsilon}{2}-\frac{A_{VF}(1+n)}{2}\right)\cdot\left(\alpha_{jn}+\alpha_{ov,n}\right)$$

$$\chi_{CSSF} = \frac{\left(1-\frac{A_{VF}(1+n)+\varepsilon}{2}\right)\left(1-\frac{R_S}{R_O}\cdot\frac{2+A_{VF}(1+n)+\varepsilon}{2}\right)\cdot\left(\alpha_{jn}+\alpha_{jp}\alpha_{np}+\alpha_{ov,n}+\alpha_{ov,p}\alpha_{np}\right)}{4\cdot\left(1-\frac{R_S}{R_O}\cdot\frac{A_{VF}(1+n)+\varepsilon}{2}\right)}$$

(5.10)

Substituting equations (5.10) and (5.9) into equation (5.8), one obtains:

$$f_{-3dB} \approx \frac{f_T}{\chi_{IN} + \chi_{OUT} + \chi_{CSSF}} = \frac{f_T}{\chi_{TOT}} \tag{5.11}$$

Equation (5.11) shows that a large f_T for the matching stage and the common source input MOST is desirable in order to increase the bandwidth. Figure 5.10 shows plots of χ_{IN}, χ_{OUT}, χ_{CSSF} and χ_{TOT} versus (ε, $g_{m2}R_S$).

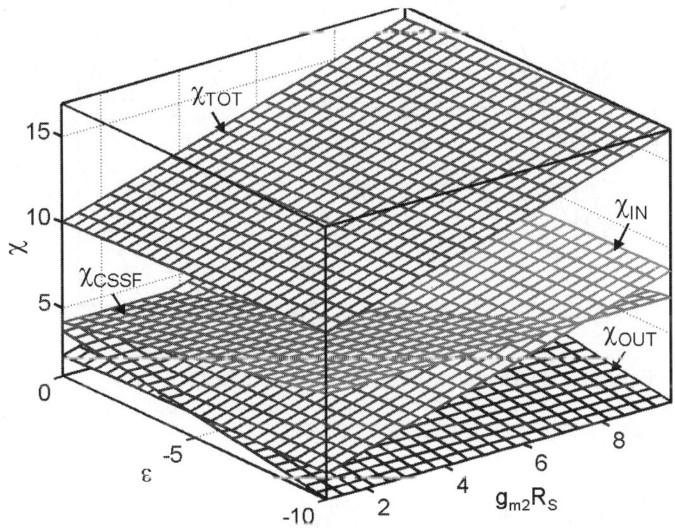

Figure 5.10: χ versus (ε, $g_{m2}R_S$) for A_{VF}=-10, n=0.2, NEF=1.5, R_O=1KΩ, L_{min}=0.25µm, $\alpha_{ov,n}$=0.2, $\alpha_{ov,p}$=0.5, α_{jn}=0.32, α_{jp}=0.28 and α_{np}=2.5.

For high $g_{m2}R_S$ and moderate $-\varepsilon$, χ_{IN} is the largest term. However, the contribution of χ_{CSSF} must not be overlooked. For a given $g_{m2}R_S$, χ_{CSSF} increases with $-\varepsilon$ because both resistance and capacitance (at this node) increase. Figure 5.11 shows the contours of the ratio f_{-3dB}/f_T versus (ε, $g_{m2}R_S$). The ratio f_{-3dB}/f_T decreases as both $g_{m2}R_S$ and $-\varepsilon$ increase because the input pole and output pole of the matching-stage degrade. For $\varepsilon=\varepsilon_{OPT}$, f_{-3dB}/f_T drops when the decrease of χ_{IN} is shadowed by the increase of χ_{CSSF}. In this case, the lower NF for $\varepsilon=\varepsilon_{OPT}$ (instead of ε=0) comes at the price of some bandwidth degradation.

From the above figures we conclude at large values of $g_{m2}R_S$ (i.e. low NF):

- The bandwidth is mainly determined by the input pole (i.e. $\chi=\chi_{IN}$) for ε=0.
- As $-\varepsilon$ increases, the output pole of the matching stage is important (i.e. $\chi=\chi_{IN}$).

- Also, $\varepsilon=\varepsilon_{OPT}$ degrades the amplifier bandwidth. However, depending on the given bandwidth specification this may not be an issue.

The above analysis of the bandwidth did not consider the use of compensation to improve the frequency response of the noise figure (see chapter 4). In this case, the bandwidth can be significantly larger than estimated because of the effect of the input pole is attenuated.

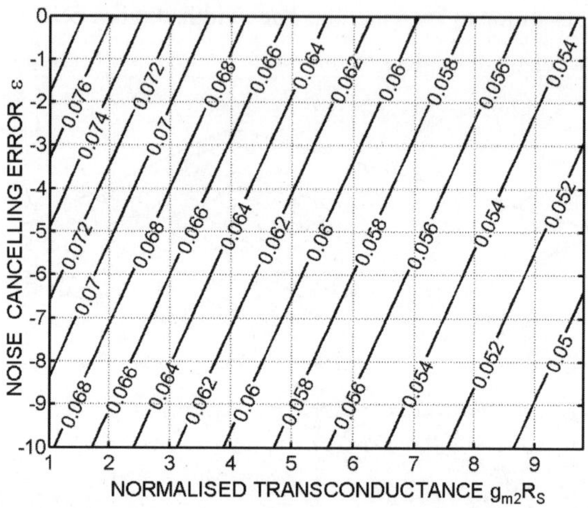

Figure 5.11: Contours of f_{-3dB}/f_T versus (ε, $g_{m2}R_S$) for A_{VF}=-10, n=0.2, NEF=1.5, R_O=1KΩ, L_{min}=0.25μm, $\alpha_{ov,n}$=0.2, $\alpha_{ov,p}$=0.5, α_{jn}=0.32, α_{jp}=0.28 and α_{np}=2.5.

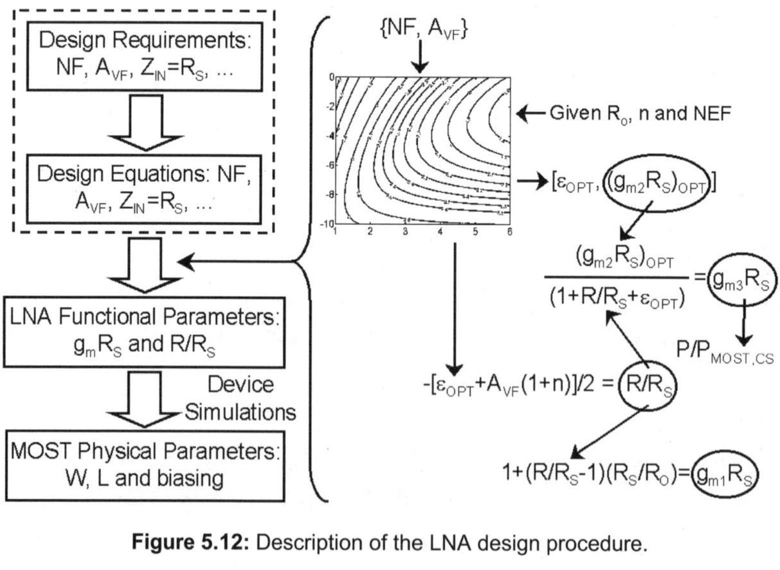

Figure 5.12: Description of the LNA design procedure.

5.4 LNA Design: NF at Minimum Power Dissipation

This paragraph deals with a design procedure to dimension the LNA in figure 5.1 in order to meet the desired noise figure NF, gain A_{VF} and $Z_{IN}=R_S$ while draining the least power P. Figure 5.12 shows the flow diagram of this procedure, which consists of the following main steps:

- Assigned NF, A_{VF} and $Z_{IN}=R_S$, and known the amplifier design equations (as given in the previous sections) the functional parameters of the LNA (i.e. g_mR_S and R/R_S) are derived according to the algorithm in the right side of figure 5.12.
- For a given technology, the physical parameters of MOSTs (i.e. W, L and biasing) are found from the previously derived functional parameters and according to basic device simulations.

The above design procedure exploits both hand-calculations and simulations in order to size the LNA. This approach is preferred, as the accuracy of simple large-signal MOST models for hand-calculations is inadequate. Table 5.1 shows the output of the procedure in figure 5.12 for NF=2dB and A_{VF}=-10, as an example. The sub-optimal design for $\varepsilon=0$ is also shown for sake of comparison.

Specs	$g_{m1}R_S$	$(g_{m2}R_S\ ,\ \varepsilon)$	$g_{m3}R_S$	R/R_S	$P/P_{MOST,CS}$	f_{-3dB}/f_T
$Z_{IN}=R_S$ NF≈2dB A_{VF}=-10	1.36	OPT: (4.6, -4.5)	1	8.2	0.5+4.6	6.3%
	1.25	SUB-OPT: (5.6, 0)	0.79	6	0.5+5.6	6.4%

Table 5.1: MOST g_mR_S's and R/R_S according to figure 5.13

Case	W_{1a} (μm)	W_{1b} (μm)	R (Ω)	W_2 (μm)	W_3 (μm)	I_{BIAS1} (mA)	I_{BIAS2} (mA)	I_{DD} (mA)
$\varepsilon=\varepsilon_{OPT}$	90	260	412	510	70	0.15	0.75	12.5
$\varepsilon=0$	80	200	300	615	50	0.18	0.85	16

Table 5.2: Device size and biasing currents for the cases $\varepsilon=\varepsilon_{OPT}$ and $\varepsilon=0$ assuming R_S=50Ω, V_{DD}=2.5V, $V_{GS}-V_{T0}$=0.25V and $L=L_{min}$=0.25μm.

Table 5.2 shows the MOSTs W/L and biasing currents as obtained from simulations with $V_{gs}-V_{T0}$≅250mV for M1a, M1b and M2a. This value of $V_{gs}-V_{T0}$ is chosen to compromise among power-efficiency, high-frequency behaviour and voltage headroom limitations. The latter issue arises from the fact that, for a given gain and I_{BIAS2}, the $V_{GS}-V_{T0}$ of the input MOS of the common-source amplifier stage must be kept relatively small in order to fit the available supply voltage (i.e. 2.5Volt). This limitation is imposed by the cascode device M2b and the diode connected load M3, which also suffers from body effect. However, a small V_{GS}-V_{T0} limits the linearity performance of the common-source stage.

5.5 Validation of the Design Procedure

The design procedure is now validated through simulations assuming the LNA sized according to table 5.2. To this purpose, we need a mean to quantify the actual level of noise cancellation occurring at the LNA output. This is done evaluating the *magnitude of the small-signal transfer function* from a test ac-current source (referred hereafter as the noise source, $I_{n,i}$) between the source and drain terminals of the matching device (M1a, for instance) to the output voltage V_{OUT}. Clearly, exact cancellation occurs when $V_{OUT}/I_{n,I}$ is zero. Figure 5.13 shows the plot of $V_{OUT,}/I_{n,i}$ versus the shunt-feedback resistor R at fixed frequencies from 1MHz to 1GHz. The LNA is sized for ε=0, C_1 and C_2 are large enough to be negligible. At lower frequencies, $V_{OUT,}/I_{n,i}$ is equal to zero for R=285Ω (i.e. exact cancellation). This value is close to R=300Ω predicted by the design procedure. As the frequency increases, the minimum of $V_{OUT,}/I_{n,i}$ rises due to C_{IN}. Figure 5.14 shows NF at f=10MHz versus R. According to figure 5.13, NF drops rapidly as R rises from 50Ω to 285Ω because more noise is cancelled at the output of the LNA. As R exceeds 285Ω, NF still drops but to a lesser extent (i.e. EF_R+EF_{CS} drops faster than the increase of EF_{MD}). Increasing R further, NF rises again (not shown) because the gain of the matching stage saturates to the intrinsic gain of the MOST while the output noise still increases.

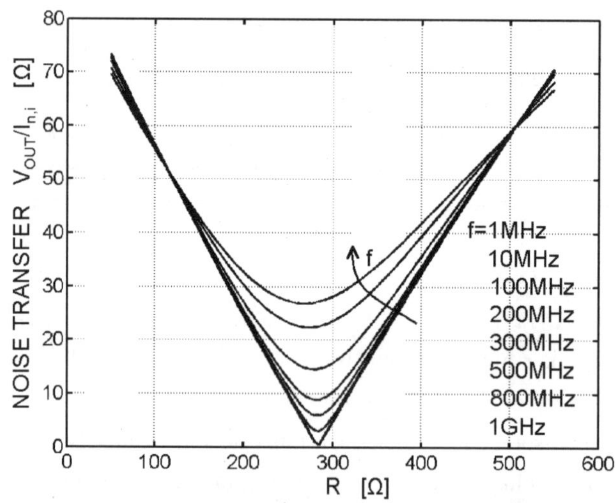

Figure 5.13: Noise transfer $V_{OUT}/I_{n,i}$ at different frequencies versus the shunt-feedback resistance R for large C_1 and C_2.

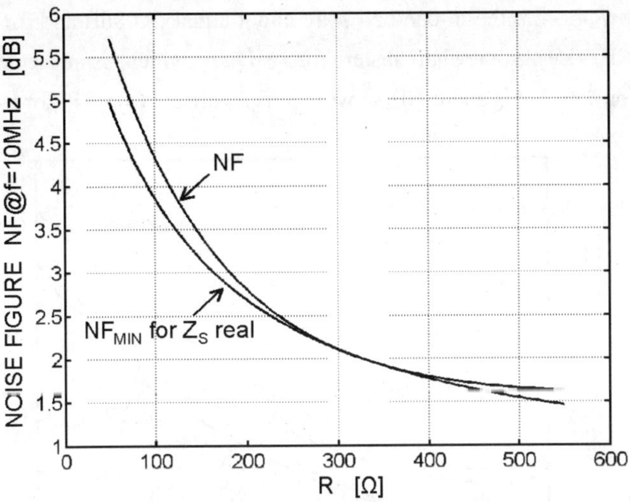

Figure 5.14: Noise figure (NF) and minimum NF_{MIN} versus R for large C_1 and C_2.

The minimum noise figure with respect to the real R_S, NF_{MIN}, is also plotted in figure 5.14. NF is equal to NF_{MIN} for R about equal to 285Ω. This validates the analysis carried out in chapter 4, where noise-power matching was shown to occur at ε=0. Figure 5.15 shows NF versus frequency for ε=0 and $\varepsilon=\varepsilon_{OPT}$=-4.5.

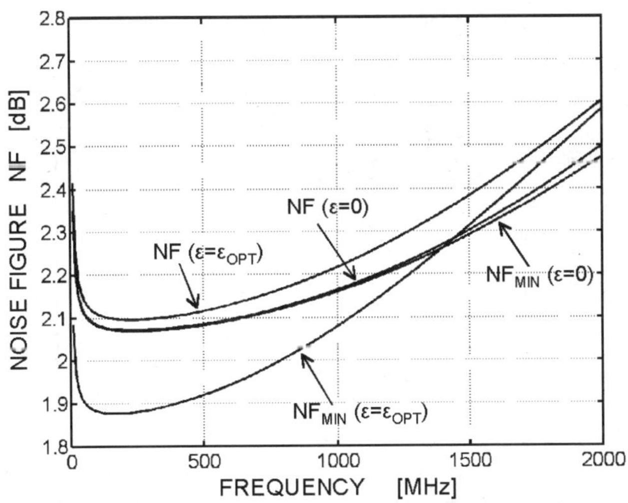

Figure 5.15: Simulated NF and NF_{MIN} versus frequency (10MHz to 2GHz) for: ε=0 and $\varepsilon=\varepsilon_{OPT}$=-4.5.

For $\varepsilon=0$ and $\varepsilon=\varepsilon_{OPT}=-4.5$, the in-band NFs are about equal (2.08dB and 2.1dB). At higher frequencies, NF rises somewhat faster for $\varepsilon=\varepsilon_{OPT}$, which is expected from hand calculations. For $\varepsilon=0$, NF is about NF_{MIN} while NF_{MIN} differs from NF for $\varepsilon=\varepsilon_{OPT}=-4.5$.

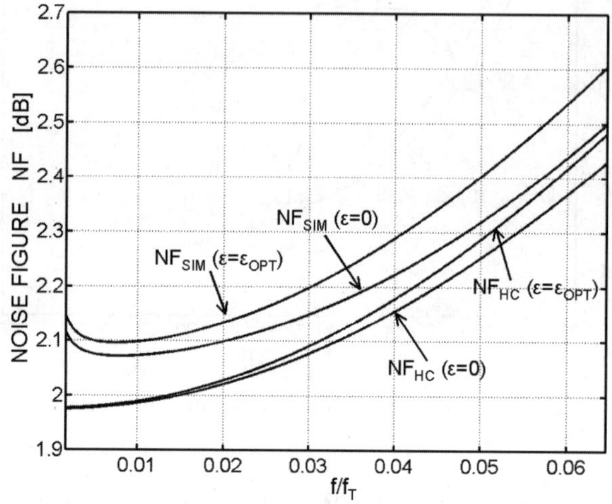

Figure 5.16: Simulated NF_{SIM} and calculated NF_{HC} versus f/f_T (f_T=31GHz @V_{GS}-V_{T0}=250mV): $\varepsilon=0$ and $\varepsilon=\varepsilon_{OPT}=-4.5$.

The simulated (MOS model 9) and hand-calculated NF curves versus f/f_T are compared in figure 5.16, showing a good agreement over the whole frequency range. Furthermore, they agree in predicting a somewhat larger high frequency degradation for $\varepsilon=\varepsilon_{OPT}=-4.5$. Figure 5.17 shows the total voltage gain versus frequency for $\varepsilon=0$ and $\varepsilon=\varepsilon_{OPT}=-4.5$.

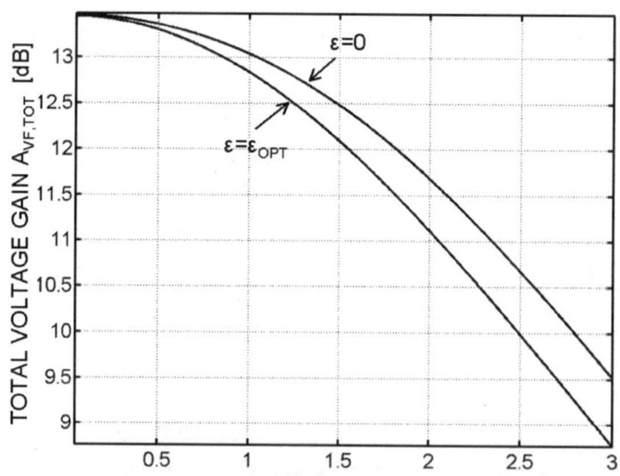

Figure 5.17: Total gain, $A_{VF,TOT}$ versus frequency for $\varepsilon=0$ and $\varepsilon=\varepsilon_{OPT}=-4.5$.

The in-band gain is 13.5dB. The bandwidth is 2.58GHz for $\varepsilon=0$ and 2.35GHz for $\varepsilon=\varepsilon_{OPT}=-4.5$, reasonably close to $f_{-3dB}\approx 6.4\% f_T=1.99$GHz ($\varepsilon=0$) and $f_{-3dB}\approx 6.3\% f_T=1.95$GHz for $\varepsilon=\varepsilon_{OPT}$ obtained from equation (5.11) for $f_T\approx 31$GHz.

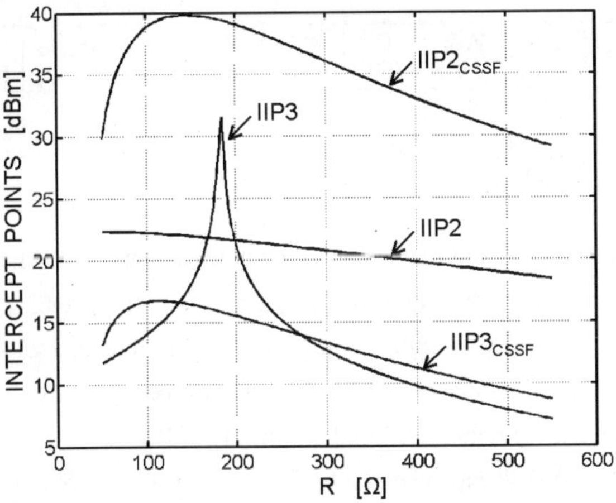

Figure 5.18: Simulated 2nd and 3rd order input-referred intercept points versus R (f_1=10MHz, f_2=12MHz) for $\varepsilon=0$ and $C_1=C_2$=large.

Figure 5.18 shows the simulated (using harmonic balance) 2nd and 3rd order input-referred intercept points versus R (design for $\varepsilon=0$) for two tones at f_1=10MHz and f_2=12MHz. Third order distortion cancellation (i.e. IIP3=∞) occurs for a value of R lower than 285Ω (i.e. 180Ω). This suggests that the distortion components generated by the common-source stage M2-M3 *subtract* to the distortion components of the matching stage M1-R (CSSF) stage. Also, noise and distortion cancelling do not occur at the same R. On the other hand, figure 5.18 does not show a peaking of IIP2 for any considered R. This suggests that second-order distortion added by the common-source stage is such that it modifies the sign relation between the two paths. For $\varepsilon=\varepsilon_{OPT}$ (i.e. R>285Ω) IIP3 further degrades.

5.6 Final Design

This paragraph deals with the final design of the LNA in figure 5.1 using the target specs indicated table 5.3. Table 5.4 shows the sizing of the LNA obtained using the design procedure outlined in figure 5.12. To do so, the following choices have been made:

Specs	$g_{m1}R_S$	$(g_{m2}R_S)_{OPT}$, ε_{OPT}	$g_{m3}R_S$	R/R_S	$P/P_{MOS,CS}$	f_{-3dB}/f_T
$Z_{IN}=R_S$ NF≈1.9dB $A_{VF}=-10$	1.35	(4.9, -4)	0.96	8	0.5+4.9	6.2%

Table 5.3: MOST g_mR_S's and R/R_S according to figure 5.12

Case	W_{1a} (μm)	W_{1b} (μm)	R (Ω)	W_2 (μm)	W_3 (μm)	I_{BIAS1} (mA)	I_{BIAS2} (mA)	I_{DD} (mA)
$\varepsilon=\varepsilon_{OPT}$	90	260	400	540	60	0.2	0.85	14

Table 5.4: Device size and biasing currents for $\varepsilon=\varepsilon_{OPT}$ assuming $R_S=50\Omega$, V_{DD}=2.5Volt, $V_{GS}-V_{T0}$=0.25Volt and $L=L_{min}$=0.25μm.

- A $C_1 \approx 13$pF guarantees a good match for frequencies as low as 50 MHz.
- The high-pass filter R2-C2 has a cut-off frequency of about 2MHz. Large R_2=95KΩ in combination with a small C_2=0.8pF are used in order to reduce area and the size of the back-plate contribution. However, noise cancellation occurs well above 2MHz due to the frequency-dependent phase-shift and amplitude of R2-C2.
- The cascode device M3 is chosen equal to the common-source NMOST, M2.

Figure 5.19 shows the result of a simple NF simulation experiment without and with noise cancellation.

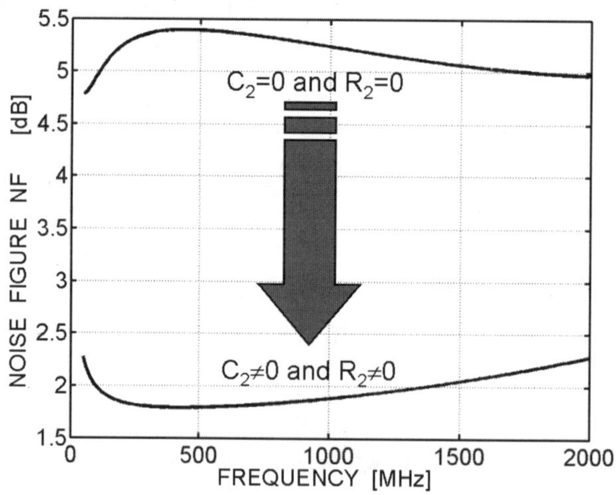

Figure 5.19: Noise figure simulation showing the existence of the noise cancelling (see text).

In the upper curve, C_2-R_2 is removed and the gate of M3 is connected to the supply. It can be seen that NF is significantly large, between 5 and 5.5dB because the LNA operates as a common-source amplifier stage with an input *active termination* of about 50Ω. As such, NF must exceed the theoretical limit $10 \cdot Log_{10}(1+NEF)>3dB$ with NEF>1. On the other hand, when C_2-R_2 is active, NF is below 2.5dB over the full range of frequencies. This can only be explained by the fact that noise cancellation takes place. Figure 5.20 shows the simulated and calculated NF versus f/f_T (f_T=31GHz@V_{GS}-V_{T0}= 250mV). Both curves are in good agreement.

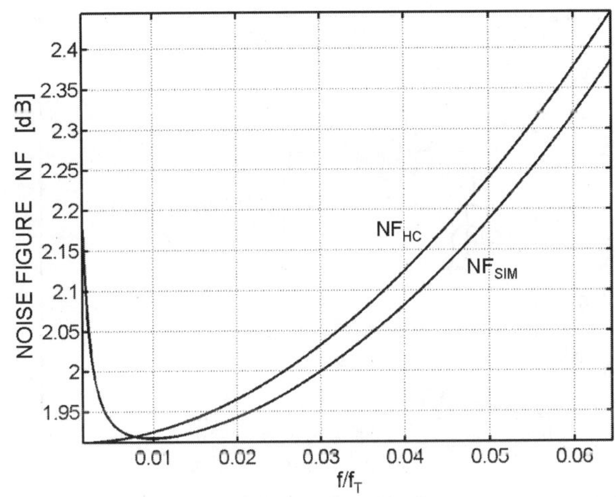

Figure 5.20: Simulated NF_{SIM} and calculated NF_{HC} versus f/f_T (f_T= 31GHz @V_{GS}-V_{T0}=250mV) for $\varepsilon=\varepsilon_{OPT}=-4$.

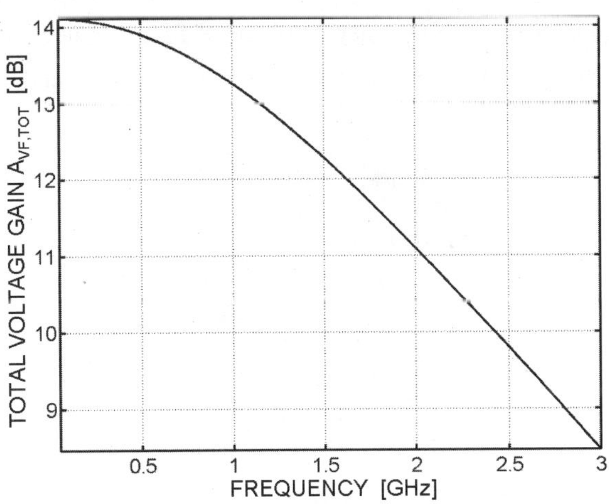

Figure 5.21: Total voltage gain $A_{VF,TOT}$ versus frequency (50MHz to 3GHz) for $\varepsilon=\varepsilon_{OPT}=-4$.

Figure 5.21 shows the total voltage gain $A_{VF,TOT}$ versus frequency. The gain is 14dB, the bandwidth is about 2.02GHz, while 6.2%f_T=1.92GHz is predicted by hand calculations. IIP3 and IIP2 were simulated using harmonic balance for two tones at (200MHz, 300MHz) and (900MHz, 905MHz). This rendered 3dBm and 14.5dBm, respectively. Since ε is non-zero, it is likely that both the matching stage and common-source stage contribute to the value of the intercepts.

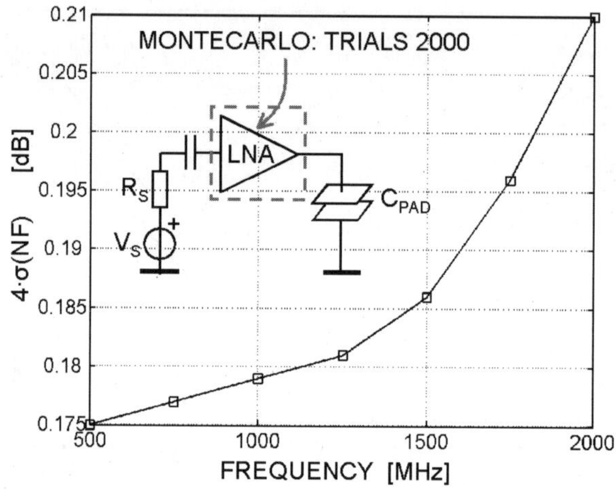

Figure 5.22: Montecarlo simulations of process-spread, 4·σ(NF), versus frequency for $\varepsilon=\varepsilon_{OPT}=-4$.

Figure 5.22 shows the sensitivity of NF@$\varepsilon=\varepsilon_{OPT}=-4$ to process-spread and mismatch. The 4·σ(NF) value is plotted versus frequency as obtained from Montecarlo simulations. NF exhibits a modest variation as 4·σ(NF) is smaller than 0.18 @1GHz and 0.22@ 2Ghz.

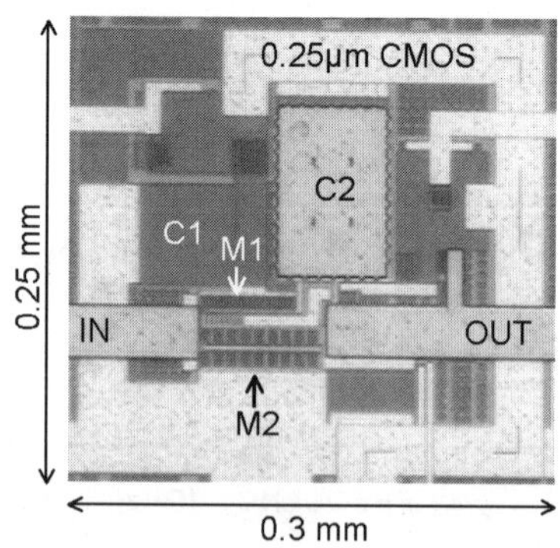

Figure 5.23: Die-photo of the wide-band 0.25μm CMOS LNA.

5.7 Measurements

The wide-band LNA in figure 5.1 has been fabricated in a standard digital 0.25μm CMOS process. The chip-photo is shown in figure 5.23.

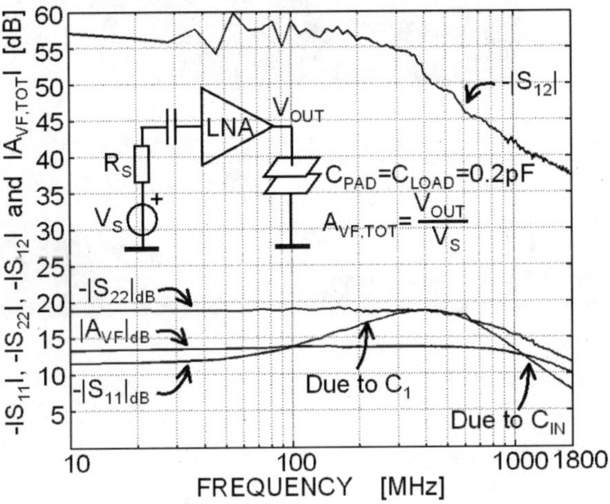

Figure 5.24: Measured (on-wafer) S_{11}, S_{22}, S_{12} and $A_{VF,TOT}$ versus frequency.

The S-parameters (i.e. S_{11}, S_{22}, S_{12} and S_{21}) of the LNA have been measured directly on wafer using an HP Vector Network Analyser (VNA). Figure 5.24 shows (the module of) S_{11}, S_{22}, S_{12} and the total voltage gain $A_{VF,TOT}$ (for C_L=0.2pF) versus the frequency from 1 to 1800 MHz. A wide band flat gain of 13.7dB is found over a 3dB bandwidth between 2 MHz and 1600 MHz. This is about 400MHz less than in simulations and it is attributed to excessive parasitic capacitance in the layout. Nevertheless, at 1800 MHz a gain of 10dB is still available. The reverse isolation $|S_{12}|$, is better than –42dB up to 1GHz and better of -36dB up to 1.8GHz. The input match, $|S_{11}|$ is better than –10dB in 10-1600 MHz and better than –8dB in 10-1800 MHz. At low frequencies, $|S_{11}|$ rises due to the shunt capacitor C_1 in the matching stage. At high frequencies, $|S_{11}|$ drops due to the input capacitance C_{IN}. Noise figure and distortion were measured with the die glued to a ceramic substrate and connected to 50Ω input/output transmission lines via short bond-wires (about 2mm). Care was taken to minimise the parasitic inductance of the ground return path by connecting several bond-wires in parallel and to de-couple properly the biasing ports.

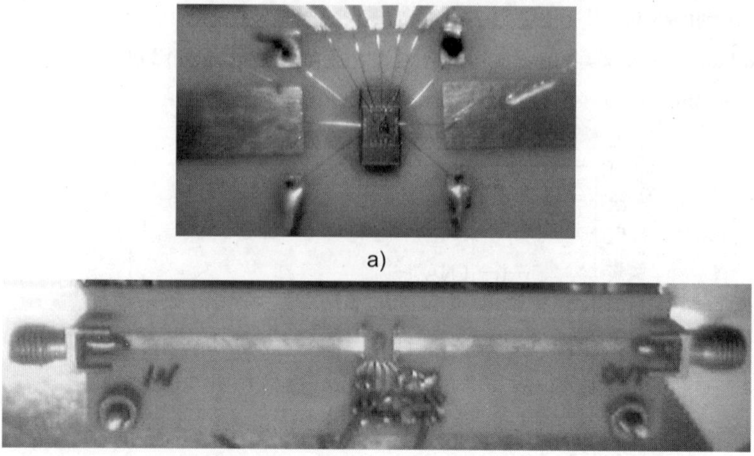

a)

b)

Figure 5.25: Photos of the PCB set-up used to measure noise figure and distortion.

Figure 5.25 shows photos of the printed circuit board (PCB). Both, the NF and IIP2 and IIP3 measurements are influenced to the short input/output bond-wires. According to simulations, a short input bond-wire has enough inductance to compensate somewhat the effect of the large input capacitance C_{IN}, thereby both the source impedance match and NF improve at high frequency. A better impedance match at the input is also beneficial in order to increase the accuracy of the NF measurement[XVIII] [4]. The noise figure measurement was performed with the PCB housed by a shielded metal-box. The PCB noise figure was measured with a calibrated HP NF-meter. Losses associated to connectors and input/output transmission lines (TLs) were measured separately on another PCB with the same TLs and connectors. Their effect was then de-embedded from measured PCB NF according to a procedure similar to that used in chapter 3. Figure 5.26 shows measured, simulated and calculated NF_{50}. The measured NF is below 2.4dB over more than one decade (150-2000 MHz) and below 2dB over more than 2 octaves (250-1100 MHz). Furthermore, the agreement with simulation and calculation is rather good.

Figures 5.27 and 5.28 show the measured IIP2 and IIP3 as obtained from extrapolated data points for two tones located at (200MHz, 300MHz) and (900MHz, 905MHz) respectively. The IIP2 is equal to +12dBm (14.5dBm in simulation) and IIP3 is equal to 0dBm (3dBm in simulations). Figure 5.29 show the 1dB compression point, which is–9dB, about 11dB

[XVIII] In this respect, a 3dB attenuator is connected to the input to improve the match of about 6dB.

below IIP3 (about 9.6dB difference is predicted by simple theory). Table 5.6 gives a complete summary of the measurements.

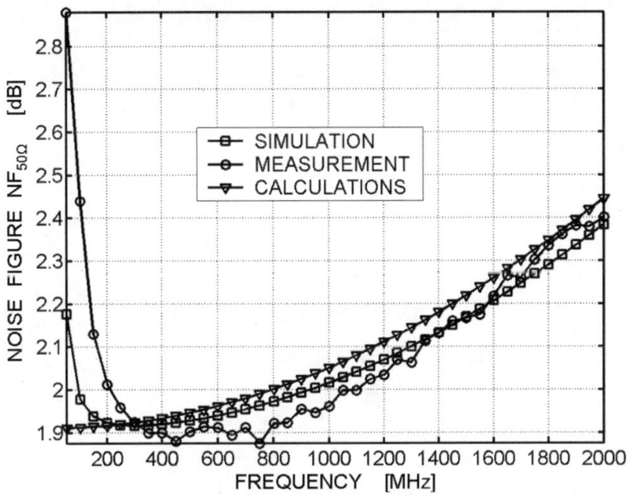

Figure 5.26: Measured, simulated and calculated NF versus frequency.

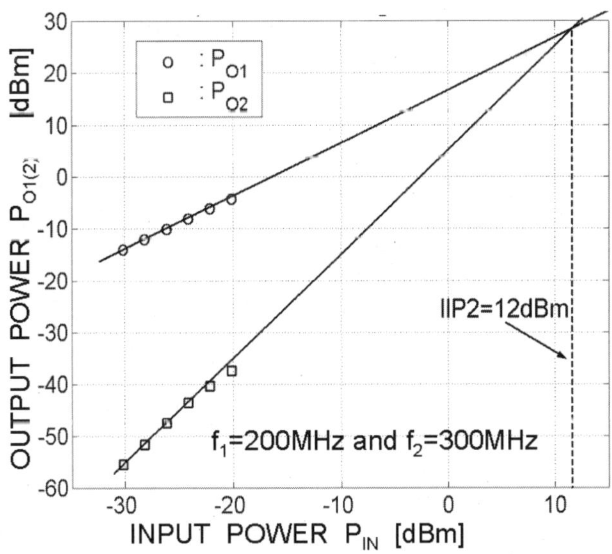

Figure 5.27: Measured (on-PCB) IIP2.

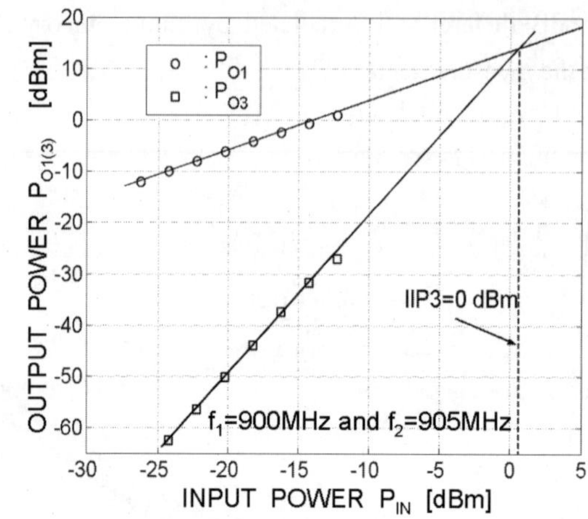

Figure 5.28: Measured (on-PCB) IIP3.

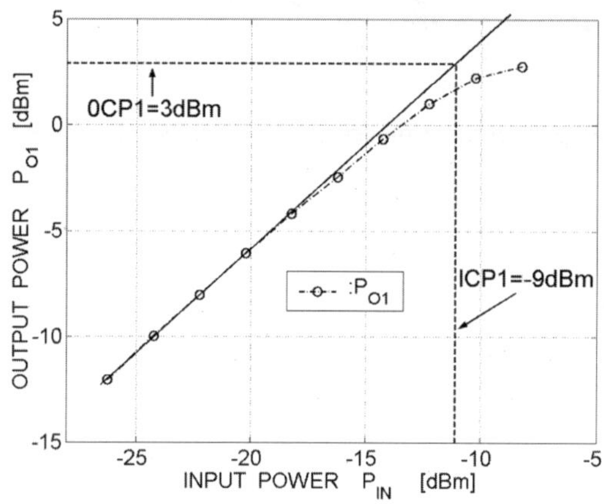

Figure 5.29: Measured (on-PCB) ICP1dB.

PROPERTY	VALUE
$\lvert A_{VF}=V_{OUT}/V_S \rvert$	13.7 dB
-3dB BW	2-1600 MHz ($C_{LOAD}=C_{PAD}$=0.2pF)
$\lvert S_{12} \rvert$	<-36dB in 10-1800 MHz
$\lvert S_{11} \rvert$	<-8dB in 10-1800 MHz
$\lvert S_{22} \rvert$	<-12dB in 10-1800 MHz
IIP3 (input ref.)	0 dBm (f_1=900MHz & f_2=905MHz)
IIP2 (input ref.)	12 dBm (f_1=200MHz & f_2=300MHz)
ICP1 (input ref.)	-9 dBm (f_1=900 MHz)
$NF_{50\Omega}$	≤2dB [0.25-1.1 GHz] & ≤2.4dB [0.15-2 GHz]
I_{DD}@V_{DD}	14mA@2.5Volt
Area and Technology	0.3x0.25mm² in a 0.25μm CMOS

Table 5.6: Summary of the LNA measurements

5.8 Summary and Conclusions

In this chapter, a wide-band low-noise amplifier has been designed in a 0.25um CMOS in order to demonstrate noise-cancelling concept.

Detailed analysis of the NF and frequency behaviour showed that:

- Exact noise cancellation does not provide the lowest noise figure. Indeed, for the same $g_{m2}R_S$ (i.e. power dissipation), voltage gain and source impedance match, a lower NF is achieved by allowing a proper noise cancellation error, $\varepsilon=\varepsilon_{OPT}<0$. Alternatively, for $\varepsilon=\varepsilon_{OPT}<0$ the same NF as for $\varepsilon=0$ is achieved at a lower power dissipation. Furthermore, NF for $\varepsilon=\varepsilon_{OPT}$ has the least sensitivity to device parameter variations.
- NF($\varepsilon=\varepsilon_{OPT}$) degrades at high frequency somewhat faster compared to NF($\varepsilon=0$). This may be acceptable. If not, frequency compensation can attenuate this effect.
- The method of the open time-constant shows that bandwidth is mainly determined by the input and output capacitance of the matching stage. Bandwidth decreases as both $g_{m2}R_S$ and $-\varepsilon$ increase. Using $\varepsilon=\varepsilon_{OPT}$, can degrade bandwidth with respect to the case $\varepsilon=0$. In this case, the lower in-band NF for $\varepsilon=\varepsilon_{OPT}$ comes at the price of some bandwidth degradation. This may be acceptable. If not, the frequency compensation exploited to improve NF can extend bandwidth too.
- LNA in figure 5.1 was designed in 0.25μm standard CMOS process aiming to high sensitivity highly-integrated receivers with demanding requirements for noise figure, voltage gain, bandwidth and input match, while draining the least current from the supply. Measurements of the demo chip showed a noise figure below 2.4dB over more than one decade (i.e. 150-2000 MHz) and below 2dB over more than 2 octaves (i.e. 250-1100 MHz). A complete summary of the measurements is shown in table 5.6.

The above experimental results prove that the feed-forward noise cancelling technique is a concrete alternative to traditional approaches relaying on global negative feedback.

5.9 References

[1] M. Steyaert and W. Sansen, "Opamp Design Towards Maximum Gain-Bandwidth", in Analog Circuit Design, edited by W. Sansen, J. Huising and R. van de Plaasche, Kluwer Academic Publishers, NY, vol.3, 1996, pp. 149-166.

[2] P. R. Gray, P. G. Hurst, S. H. Lewis and R. G. Meyer, "Analysis and Design of Analog Integrated Circuits", John Wiley & Sons, 4rd edition.

[3] T. H. Lee, "The Design of Radio-Frequency CMOS Integrated Circuits", Cambridge University Press, 1998.

[4] H. Packard, "Noise Figure Measurement Accuracy", Application Note 57-2, November 1988.

[5] P. Wambacq and W. M. Sansen, “Distortion Analysis of Analog Integrated Circuits”, Kluwer Academic Publisher, July 1998.

Chapter 6
Summary and Conclusions

6.1 Summary and Conclusions

In this book, high-performance wide-band low-noise amplifiers (LNAs) suitable for low-cost standard CMOS processes have been investigated. The focus was on low-noise techniques suitable for communication receiver applications with potentially demanding requirements on the LNA performance such as noise figure (NF) well below 3dB, sufficiently large gain and source impedance matching. Next, good linearity and the possibility for some variable gain are desired in order to handle the interference generated by strong adjacent channels. Moreover, the previous requirements must be met over a range of frequencies that can exceed one octave or even a decade depending on the specific application.

In order to find amplifier topologies that can meet these requirements, designers generally relay on their creativity, insight and experience. As this process is a largely unstructured, it is unlikely that *all* the potentially useful amplifier alternatives are found. In this book, the design problem is faced using a radically different approach described in chapter 2. A*ll* the potentially useful wide-band amplifiers are investigated by applying a methodology that generates systematically *all* the two-port amplifiers that can be modelled as circuits with 2 Voltage Controlled Current Sources (VCCS). The choice of two VCCS as generating elements is motivated by the following facts:

- Commonly used elementary amplifiers [1] as the common gate, common drain and the common source shunt-feedback stages exploit the "g_m" of at least 1 or two MOSTs to define their small-signal transfer properties as gain, port impedance and bandwidth. These circuits can be regarded as circuits with 1 VCCS or 2 VCCSs.
- In previous work [2,3,4], all the graphs of potentially useful two-port circuits built by the interconnection of 2 VCCSs were systematically generated, classified in terms of their properties and stored in a database. As such, all the graphs of 2VCCS two-port wide-band amplifiers are contained in the 2VCCS database.

Elementary implementations of the generated 2VCCS amplifiers using a single-MOST (or a resistor) for each VCCS are shown in figure 6.1. Two of them (i.e. A2 and A4) are well-known circuits while the other two (i.e. A1 and A3), at the best of our knowledge, are new topologies of wide-band amplifiers.

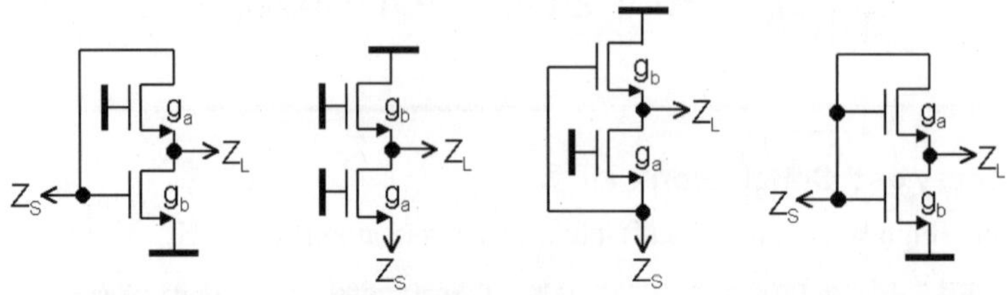

Figure 6.1: Systematically generated 2-MOSFET wide-band amplifiers (biasing not shown).

In chapter 3, important aspects of the performance of these amplifiers are analysed in order to find out whether the new amplifiers A1 and A3 can perform better with respect to A2 and A4. Using simple modelling, hand calculations showed that:

- The gain A_{VF} of A1 is larger than that of A4 for the same value of g_a and g_b. For the same A_{VF}, C_L and Z_{IN}=R_S, the output impedance of A4 is larger leading to a somewhat lower expected bandwidth. The noise factor of A1 is better than that of A4 due to its lower output noise for the same gain A_{VF} and Z_{IN}=R_S.
- The gain A_{VF} of A3 is larger than that of A2 (and amplifier A4) for the same value of g_a and g_b. For the same A_{VF}, C_L and Z_{IN}=R_S, the output impedance of A2 is larger leading to a lower expected bandwidth.
- Amplifier A3 shows a gain-independent F=1+NEF for Z_{IN}=R_S and NEF_a=NEF_b=NEF. This value is the lowest among the amplifiers in figure 6.1 regardless the gain and for equal power. For g_bR_S=1 (A_{VF}=2), *cancellation of the output noise from the matching device* occurs regardless the value of Z_{IN}. In this case, the limitation F=1+NEF arises because the noise cancelling condition constraints the g_b of the upper MOST to 1/R_S.

To exploit the properties of the amplifier A3, the wide-band LNA in figure 6.2 was designed in a 0.35μm digital CMOS. Measured and simulated NF at 500MHz is shown in figure 6.3 versus A_{VF}. The simulated (MOS model 9) NF of a common-gate amplifier is shown for comparison. Figure 6.3 shows a more or less constant NF at least 2dB better than for the CG upon the same A_{VF}, Z_{IN} and power. A summary of the measurements at maximum gain is shown in table 6.1. This proves that the systematic generation methodology can lead to new amplifier topologies with useful properties. The noise cancellation mechanism occurring in amplifier A3 is a good example.

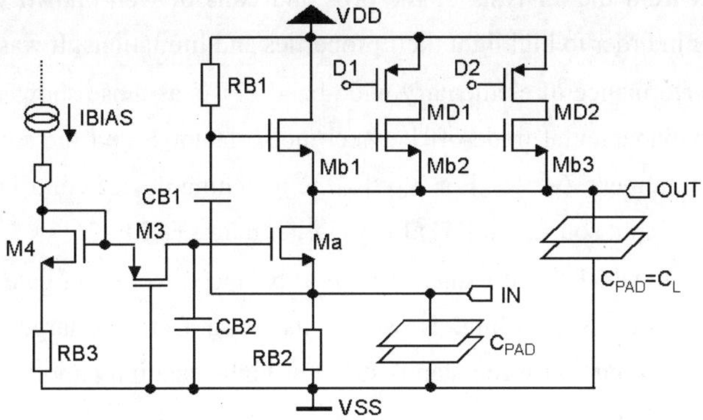

Figure 6.2: Schematic of a wide-band LNA based on the amplifier A3 (the output bond pad is used as load capacitance).

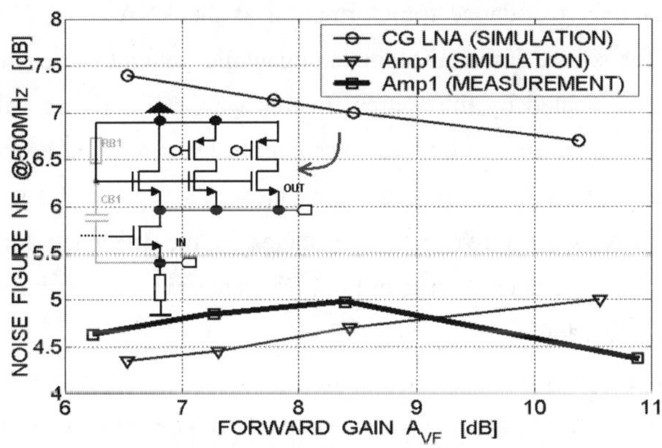

Figure 6.3: NF_{50}@500MHz versus A_{VF} for the LNA of figure 6.2 and a CG amplifier derived from it (see text).

PROPERTY	VALUE
$\lvert A_{VF}=V_{OUT}/V_{IN}\rvert$	11 dB
-3dB BW	1-900MHz (C_L=0.28pF)
$\lvert A_{VR}=V_{IN}/V_{OUT}\rvert$	< -30 dB up to 900 MHz
$VSWR_{IN}$	< 1.6 up to 900 MHz
IIP3 (input ref.)	14.7 dBm
IIP2 (input ref.)	27.4 dBm
ICP1 (input ref.)	-6 dBm
$NF_{50\Omega}$	< 4.4 dB
I_{DD}@V_{DD}	1.5mA@3.3Volt
Technology & Die area	0.35µm CMOS & 0.06 mm^2

Table 6.1: Measurements at maximum gain

Chapter 4 starts from the analysis of the pros and cons of well-known wide-band low-noise techniques in order to highlight their properties and limitations. It was found that:

- The noise performance of elementary wide-band LNAs as those shown in figure 6.1 is limited by a fundamental trade-off between noise factor F and the source impedance matching requirement $Z_{IN}=R_S$. For $Z_{IN}=R_S$ the matching device contribute to F at least as much as the input source, so F is always larger than 1+NEF>2.
- A somewhat lower F is obtained using a balanced common-gate amplifier and capacitive input cross coupling. However, its noise factor is larger than 1+NEF/2 because the matching constraint stands still. For high-sensitivity applications, 1+NEF/2 is just not enough low.
- Practical wide-band LNAs exploiting lossy negative feedback can break the trade-off between F and $Z_{IN}=R_S$ without degrading the quality of the source match. This is done exploiting an active impedance transformation at the LNA input (e.g.: figure 4.6a), which allows for source impedance matching while generating less *equivalent* input noise than that generated by the source. In this way, noise figure values well below 3dB are possible at the price of power dissipation. Despite its noise performance, the use of negative feedback has some drawbacks too:
  - ❖ Wide-band LNAs with two or more stages in the feedback loop are prone to instability. This is especially true at high frequency where the phase shit introduced by the poles is larger.
  - ❖ The input impedance Z_{IN} depends on all the circuit parameters (e.g: $g_{m,Av}$, $R_{o,Av}$ and $G_{m,I}$ for figure 4.6a), thus: (a) Z_{IN} is sensitive to device parameter variations and (b) Since Z_{IN} and A_{VF} are directly coupled, variable gain at $Z_{IN}=R_S$ requires complex circuitry.
  - ❖ Furthermore, good linearity is subordinated to the availability of a sufficiently large loop gain. The latter is typically scarce at RF frequency OR it may lead to conflicting requirements. For instance, the linearity of the commonly used feedback LNA in figure 4.6a can be as poor as that of its loop amplifier Av. This is because, regardless the gain of the loop amplifier Av, the loop-gain for $Z_{IN}=R_S$ is always < 1. When amplifier Av can consist of two or more cascaded stages with most of the gain in the first one (i.e. for best noise performance), the overall linearity can be poor [4]. In fact, for this topology another trade-off exist between $Z_{IN}=R_S$ and distortion performance.

To overcome the above limitations a novel wide-band low-noise technique was conceived, which is an evolution of the noise cancelling mechanism found for amplifier A3 in figure 6.1. The basic idea behind noise cancelling is that the trade-off between F and $Z_{IN}=R_S$ can be broken by cancelling the output noise from the matching device without degrading signal transfer and source match. This is possible because two nodes (X and Y) of the matching amplifier stage can be found where, due to their *correlation in sign*, the noise of the matching device and signal can be *distinguished and processed differently.* For instance using the matching stage in figure 6.4, the instantaneous noise voltages at node X and Y have *equal sign* while signal voltages at the same nodes have *opposite sign.*

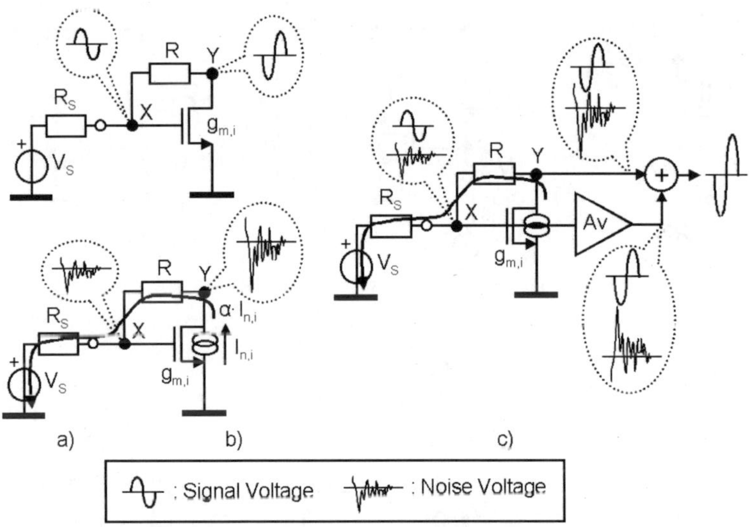

Figure 6.4: Noise cancelling principle applied to a common-source shunt-feedback stage

This means that by defining two feed-forward paths from nodes X and Y to a new output as shown in figure 6.4c, the noise contribution from the matching device can be cancelled while adding in phase the signals. This is done exploiting a proper voltage-sensing auxiliary amplifier Av connected to the output of the matching stage in feed-forward. Now, the noise figure of the noise cancelling LNA can be lowered well below 3dB by decreasing the EF of the "unmatched" auxiliary amplifier Av at the price of extra power dissipation (so as for wideband LNAs exploiting negative feedback). Note that any undesired signal or noise contribution that can be modelled as a current source between the drain and source terminal of the matching device is cancelled as well (e.g.: MOST 1/f noise, thermal noise of the distributed resistance of the gate and biasing noise).

This noise cancelling was generalised to other circuit implementations according to a simple model represented by two two-port amplifiers connected in feed-forward. One two-port provides the LNA input impedance and the other one senses the matching device noise (and signal) voltage across R_S without loading it. The outputs of the two two-ports are then properly combined in order for noise cancelling to take place at the output while adding signals. According to the two-port model, different noise-cancelling LNA circuits have been systematically generated (figure 6.5).

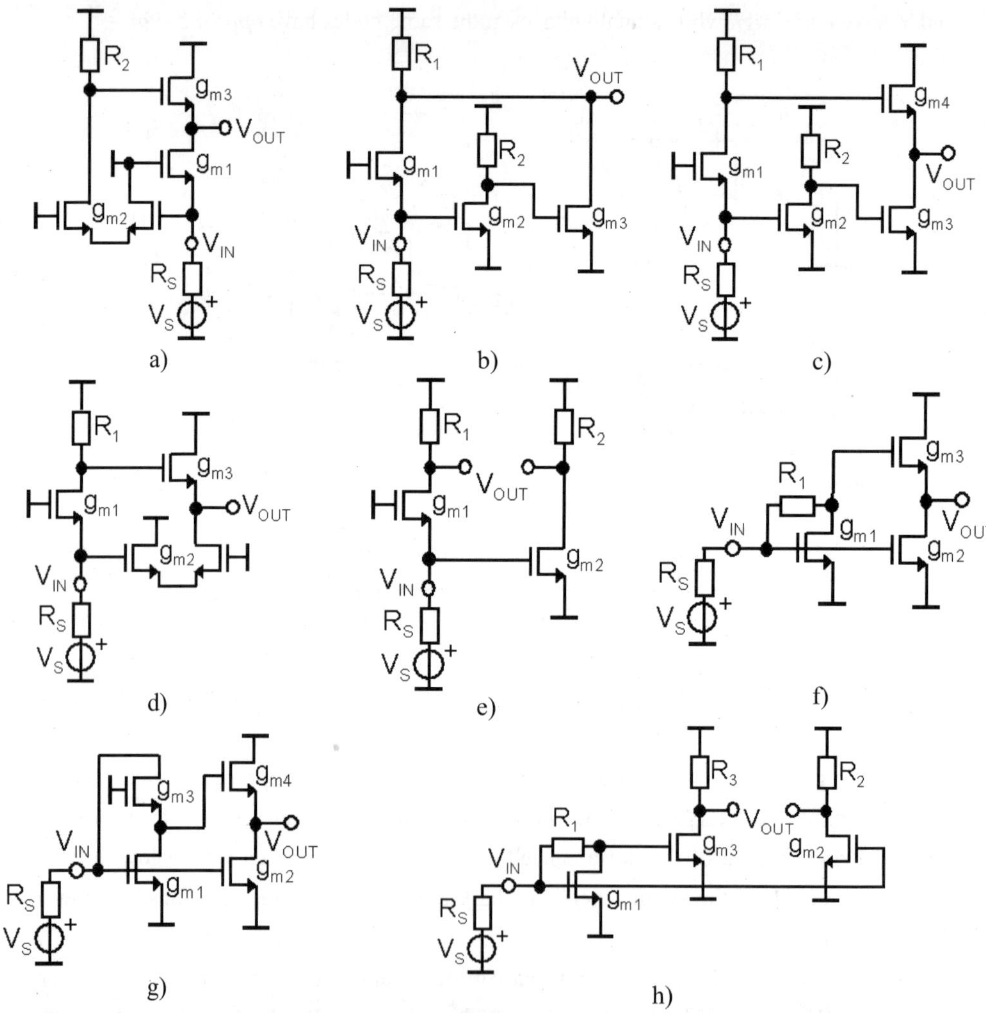

Figure 6.5: systematically generated alternative noise-cancelling amplifier (biasing not shown).

The noise performance of these LNAs was compared based on hand calculations. For $Z_{IN}=R_S$, LNA-f renders the lowest F, for a given power, gain A_{VF}, including the noise from the biasing circuitry of the matching stage and supply voltage limitations. This is because LNA-f uses a minimum of MOSTs/resistors (i.e. least noise sources and good power efficiency) and noise from the matching-stage biasing cancels at the output too. Furthermore, this LNA is also suited for high frequencies because the local shunt-feedback in the matching stage provides lower output impedance compared to the CG stage for the same load resistor and node capacitance.

Peculiar properties of noise cancelling have been also identified:

- Noise cancelling is a feed-forward technique, so instability risks are relaxed.
- Noise cancellation relies on the value of the real resistance of the source, R_S.
- Noise cancelling is *robust* to device parameter variations because the cancellation depends on a *reduced-set* of circuit parameters and variations of the source impedance RS and the gain Av of the auxiliary amplifier "Av" render a modest increase of the matching device EF_i.
- The input impedance Z_{IN} depends only on the matching device g_m. This means that:
  - ✓ Z_{IN} is less sensitive to device parameter variations than for the feedback LNA.
  - ✓ Variable gain at constant match $Z_{IN}=R_S$ is easier implemented (e.g.: in figure 6.5b, by varying the value resistor R_o).
- Noise cancelling allows for (in-band) *simultaneous noise and power* matching without an extra wideband matching network. This occurs because, upon noise cancellation, the noise transfer functions of all devices remain equal upon short and open conditions at the input of the LNA. This is strictly true for a MOST that is regarded as a VCCS.
- Assuming a *linear* voltage-sensing auxiliary amplifier, all the output distortion components generated by the matching-device are cancelled in the same manner and upon the same conditions noise cancels at the output of the LNA. In other terms, *simultaneous matching-device noise and distortion cancellation* is achieved. When the non-linearity of the auxiliary amplifier is introduced, different scenarios are possible depending on how its distortion components do combine at the output with that due to the matching-device. Under proper conditions, total cancellation of even *or* odd output distortion still occurs but this will not happen simultaneously with noise cancellation.

High-frequency limitations to noise cancelling have been also investigated. It was found that the effective input capacitance of a noise-cancelling LNA is the main reason for the degradation of the cancellation and the consequent increase of the noise figure. This occurs because the matching device noise current sees frequency dependent source impedance that affects the two feed-forward noise transfers in a different manner. In the same manner, this input capacitance is also responsible for the degradation of the distortion cancellation as well the simultaneous noise and power matching. Some design guidelines has been given in order to mitigate the noise figure degradation at high frequency. One can design the LNA for minimum parasitic input capacitance using a larger g_m/I_D for the matching device, cascoding in the auxiliary amplifier to reduce the Miller effect, design ESD protection device and use an input bond-pad with low parasitic capacitance and a more advanced sub-micron CMOS process with a higher f_T. If the previous measures are ineffective, a frequency-compensation network can be used. In this respect, two options have been examined with regard to a specific LNA topology: (a) capacitive noise equalization and (b) peaking networks. The former neutralizes the frequency dependence of EF_{MD} by equalizing the effect that the complex source impedance has over the noise transfer function of the two feed-forward paths. However, the frequency-independence of EF_{MD} comes at the price of a significant degradation of the EF of the auxiliary amplifier. The net result is than an even larger LNA noise figure. Furthermore, the equalization capacitance degrades gain and matching at high frequency. More effectively, peaking coils can be used to mitigate the degradation of noise cancelling by compensating, to some extent, the frequency-dependence of the effective source impedance seen by the noise current flowing out from the matching device. The great advantage of peaking networks is that they improve the frequency dependence of *all* devices EF, which always results in an improvement of the LNA noise figure. Furthermore, the frequency response of the gain and matching versus frequency are improved too. More specifically, two cases of peaking networks have been examined: a shunt peaking coil in series to R_S and a properly connected bridged T-coil network. The former is simple to use but yields a more limited NF improvement versus frequency whereas the latter is only somewhat more complex and renders superior results.

In chapter 5, the noise cancelling theory is validated by design. To this purpose, the noise cancelling topology of figure 6.5f was selected because of its expected superior NF performance and good high frequency capabilities. In order to optimise the LNA noise performance, the behaviour of NF and bandwidth were analysed taking into account circuit details. It was found that:

- Exact noise cancellation does not provide the lowest noise figure. Indeed, for the same $g_{m2}R_S$ (i.e. power dissipation), voltage gain and source impedance match, a lower NF is achieved by allowing a proper noise cancellation error, $\varepsilon=\varepsilon_{OPT}<0$. Alternatively, for $\varepsilon=\varepsilon_{OPT}<0$ the same NF as for $\varepsilon=0$ is achieved at a lower power dissipation. Furthermore, NF for $\varepsilon=\varepsilon_{OPT}$ has the least sensitivity to device parameter variations.

- NF($\varepsilon=\varepsilon_{OPT}$) degrades at high frequency somewhat faster compared to NF($\varepsilon=0$). This may be acceptable. If not, frequency compensation can attenuate this effect.
- The method of the open time-constant shows that bandwidth is mainly determined by the input and output capacitance of the matching stage. Bandwidth decreases as both $g_{m2}R_S$ and $-\varepsilon$ increase. Using $\varepsilon=\varepsilon_{OPT}$, can degrade bandwidth with respect to the case $\varepsilon=0$. In this case, the lower in-band NF for $\varepsilon=\varepsilon_{OPT}$ comes at the price of some bandwidth degradation. This may be acceptable. If not, the frequency compensation exploited to improve NF can extend bandwidth too.

The LNA in figure 6.6 was designed in 0.25μm standard CMOS process aiming to high sensitivity highly-integrated receivers with demanding requirements for noise figure, voltage gain, bandwidth and input match, while draining the least current from the supply. Its design was targeted for a sub-2dB noise figure, 20dB voltage gain and wide operation bandwidth. Figure 6.7 shows the plot of the 50Ω noise figure measurement. At lower frequencies, NF increases due to the HPF C_2-R_2, at high frequency the rise is due to C_{IN}. Anyhow, the noise figure is below 2.4dB over more than one decade (i.e. 150-2000 MHz) and below 2dB over more than 2 octaves (i.e. 250-1100 MHz). A summary of the measurements is given in Table 6.2. These experimental results prove that the (feed-forward) noise-cancelling concept is a valid alternative to traditional negative feedback techniques for the design of high-performance wide-band low-noise CMOS amplifiers.

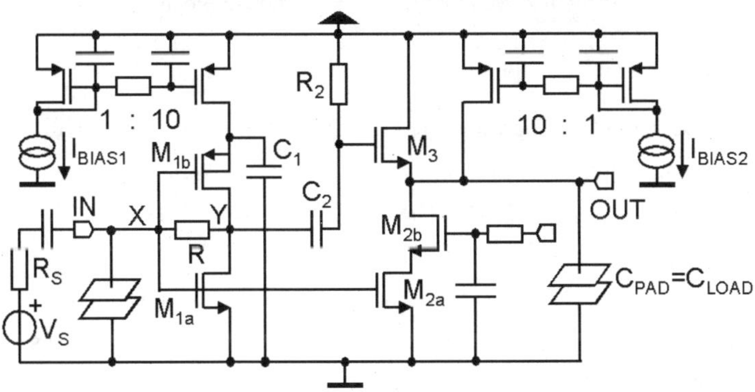

Figure 6.6: Complete schematic of the designed wide-band noise cancelling LNA.

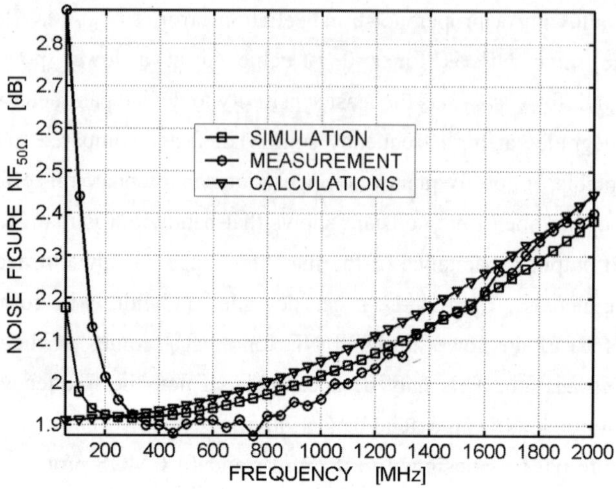

Figure 6.7: Measured and simulated NF versus frequency.

PROPERTY	VALUE
$\lvert A_{VF}=V_{OUT}/V_S \rvert$	13.7 dB
-3dB BW	2-1600 MHz ($C_{LOAD}=C_{PAD}=0.2pF$)
$\lvert S_{12} \rvert$	<-36dB in 10-1800 MHz
$\lvert S_{11} \rvert$	<-8dB in 10-1800 MHz
$\lvert S_{22} \rvert$	<-12dB in 10-1800 MHz
IIP3 (input ref.)	0 dBm (f_1=900MHz & f_2=905MHz)
IIP2 (input ref.)	12 dBm (f_1=200MHz & f_2=300MHz)
ICP1 (input ref.)	-9 dBm (f_1=900 MHz)
$NF_{50\Omega}$	≤2dB [0.25-1.1 GHz] & ≤2.4dB [0.15-2 GHz]
I_{DD}@V_{DD}	14mA@2.5Volt
Area and Technology	0.3x0.25mm^2 in a 0.25μm CMOS

Table 6.2: Summary of the LNA measurements

6.2 References

[1] B. Razavi, "Design of Analog Integrated CMOS Circuits", Mc-Graw Hill, 2001.

[2] E.A.M. Klumperink, "Transconductance based CMOS Circuits: Generation, Classification and Analysis", PhD. Thesis, University of Twente, Enschede, The Netherlands, 1997.

[3] E.A.M. Klumperink at al., "Finding all the elementary circuits exploiting transconductance", Proc. IEEE International Symposium on Circuits and Systems, Vol. I, pp. 667-670, 6-9 may 2001, Sydney, Australia.

[4] E.A.M. Klumperink at al, "Finding all the elementary circuits exploiting transconductance", IEEE Transaction on Circuits and Systems, CAS-II.

Appendix A
Two-port Amplifiers Stability and {A, B, C, D} Parameters

In the design of RF and microwave amplifiers, it is a common practice to require its stability to be unconditional [1]. The latter, means a stable operation for any value of the passive source and load terminations. Necessary and sufficient conditions for the unconditional stability of a linear two-port circuit are:

$$\Re\{Z_{IN}\} > 0 \quad \text{and} \quad \Re\{Z_{OUT}\} > 0 \qquad \forall\, Z_L \quad \forall\, \omega \tag{A.1}$$

where $\Re$ is the real part of $\{\cdot\}$. Conditions (A.1) can be shown to be equivalent to [2]:

$$\Re\{Z_{IN}\} > 0 \quad \text{and} \quad \Re\{Z_{22}\} > 0 \qquad \forall\, Z_L \quad \forall\, \omega \tag{A.2}$$

Relations (A.2) can be rewritten in terms of {A, B, C, D} parameters as:

$$\Re\{Z_{IN}\} = \Re\left\{\frac{Z_L A + B}{Z_L C + D}\right\} > 0 \quad \text{and} \quad \Re\{Z_{22}\} - \frac{D}{C} > 0 \qquad \forall Z_L \quad \forall\omega \tag{A.3}$$

Substituting $Z_L = u(\omega) + j\cdot r(\omega)$ with $u(\omega) \geq 0$ into relations (A.3), one obtains:

$$\Re\{Z_{IN}\} = \frac{[u(\omega)\cdot A + B]\cdot[u(\omega)\cdot C + D] + r(\omega)^2\cdot A\cdot C}{[u(\omega)\cdot C + D]^2 + r(\omega)^2\cdot C^2} > 0$$

$$\Re\{Z_{22}\} = \frac{D}{C} > 0 \;\; \forall r,\; \forall u > 0,\; \forall\omega$$

and so:

$$\begin{aligned} & A\cdot C\cdot\left[u(\omega)^2 + r(\omega)^2\right] + u(\omega)\cdot[C\cdot B + A\cdot D] + B\cdot D > 0 \\ & \frac{D}{C} > 0 \quad \forall r,\; \forall u > 0,\; \forall\omega \end{aligned} \tag{A.4}$$

From simple reasoning[XIX], it can be easily verified that necessary and sufficient conditions to meet relation (A.4) are that all the {A, B, C, D} parameters share the *same sign*. In our case, the load is an on-chip capacitance whose value is reasonably well defined. In such a case, stability can be ensured using $\Re\{Z_{IN}\}>0$ and $\Re\{Z_{OUT}\}>0$ for $\forall\omega$ as:

$$\begin{aligned} \Re\{Z_{IN}\} &= \frac{[u(\omega)\cdot A+B]\cdot[u(\omega)\cdot C+D]+r(\omega)^2\cdot A\cdot C}{[u(\omega)\cdot C+D]^2+r(\omega)^2\cdot C^2} > 0 \\ \Re\{Z_{OUT}\} &= \frac{[m(\omega)\cdot D+B]\cdot[m(\omega)\cdot C+A]+n(\omega)^2\cdot D\cdot C}{[m(\omega)\cdot C+A]^2+n(\omega)^2\cdot C^2} > 0, \quad \forall\omega \end{aligned} \tag{A.5}$$

where $Z_S=m(\omega)+j\cdot n(\omega)$ has been substituted and $m(\omega)\geq 0$ holds. Substituting $\{u(\omega)=0, r(\omega)=-1/(\omega C_L)\}$ and $\{m(\omega)=R_S, n(\omega)=0\}$ into relation (A.5), one obtains:

$$R_S^2\cdot D\cdot C+R_S\cdot[B\cdot C+A\cdot D]+A\cdot B>0, \quad \forall\omega \tag{A.6-a}$$

$$\frac{A\cdot C}{\omega^2 C_L^2}+B\cdot D>0 \quad \forall\omega \tag{A.6-b}$$

Again, if the {A, B, C, D} parameters share the same sign relations (A.6-a) and (A.6-b) are fulfilled. To verify that this condition is also necessary proceed as follows. Assuming for instance $\omega\rightarrow 0$, relation (A.6-b) is true if parameters "A" and "C" share the same sign. For $\omega\rightarrow\infty$ parameters "B" and "D" must share the same sign. To prove this statement, it is sufficient to show that one of parameters "A" and "C" shares the same sign with one of parameters "B" or "D". To do so, from relation (A.6-a) we observe that if this would not be the case, all the product among the {A, B, C, D} parameters render negative values (i.e. relation (A.6-a) is violated for any value of ω).

References

[1] G. Gonzalez, "Microwave Transistor Amplifiers" Prentice Hall, 2nd edition, 1984.

[2] M. Ohtomo, "Proviso on the Unconditional Stability Criteria of Linear Two-port" IEEE Transaction On Microwaves Theory and Techniques, vol. 43, No 5, pp. 1197-1200, May 1995.

[XIX] For instance evaluating relations (A.4) for $Z_L=0$ and $Z_L=\infty$.

Appendix B
Biasing noise in Noise-cancelling Amplifiers

In this appendix, expressions of the noise factor, F_{BIAS}, due to noise in the matching-stage bias circuitry are derived for the noise-cancelling amplifiers of figure 4.16. The aim is to quantify the impact of F_{BIAS} on F as a function of the parameters A_{VF}, V_{DD}, V_{GT1} and θ.

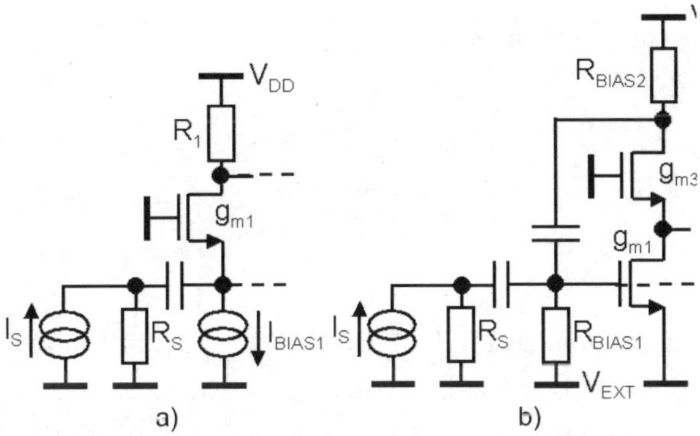

Figure 1: Biasing of the matching stage used in the LNAs of figure 4.16.

For the matching stages shown in figure 1, F_{BIAS} can be written as:

$$F_{BIAS} = 1 + \begin{cases} NEF \cdot g_{BIAS1} \cdot R_S \\ \dfrac{R_S}{R_{BIAS2}} \end{cases} \tag{B.1}$$

where the noise of R_{BIAS1} is neglected because its value can be large. Multiplying for the drain current I_D and using the matching condition $g_{m1}=1/R_S$, equation (B.1) becomes:

$$F_{BIAS} = \begin{cases} 1 + NEF \cdot \dfrac{\dfrac{g_{BIAS1}}{I_D}}{\dfrac{g_{m1}}{I_D}} \\ 1 + \dfrac{1}{V_{BIAS2} \cdot \dfrac{g_{m1}}{I_D}} \end{cases} \tag{B.2}$$

where V_{BIAS2} is the bias voltage across R_{BIAS2}. The latter, according to a simple strong inversion model (i.e. $I_D=K\cdot V_{GT}^2/(1+\theta\cdot V_{GT})$ with $V_{GT}=V_{GS}-V_{T0}$), can be written as:

$$\frac{g_{m1}}{I_D}=\frac{2}{V_{GT1}}\cdot\frac{1+\frac{\theta}{2}\cdot V_{GT1}}{1+\theta\cdot V_{GT1}}$$
$$\frac{g_{BIAS1}}{I_D}=\frac{2}{V_{GT,BIAS1}}\cdot\frac{1+\frac{\theta}{2}\cdot V_{GT,BIAS1}}{1+\theta\cdot V_{GT,BIAS1}} \tag{B.3}$$

From circuit inspection one can write:

$$V_{BIAS2}=V_{DD}-V_{DS3}-V_{DS1}$$
$$V_{BIAS1}=V_{DD}-V_{DS1}-\frac{A_{VF}}{2\cdot\frac{g_{m1}}{I_D}(V_{GT1},\theta)} \tag{B.4}$$

A_{VF} is the LNA gain. From (B.2), (B.3) and (B.4), F_{BIAS} is minimized when the voltage across the biasing device is maximized. This occurs when the MOSTs in the signal path are at the edge of saturation for $V_{DS3}=V_{GT3}$ and $V_{DS1}=V_{GT1}$ in figure 1-b and for $V_{DS1}=V_{GT1}$ in figure 1-a, which yields:

$$V_{BIAS2}=V_{DD}-V_{GT3}-V_{GT1}$$
$$A_{VF}=2\cdot\frac{V_{GT3}}{V_{GT1}}\cdot\frac{1+\frac{\theta}{2}\cdot V_{GT1}}{1+\theta\cdot V_{GT1}}\cdot\frac{1+\theta\cdot V_{GT3}}{1+\frac{\theta}{2}\cdot V_{GT3}} \tag{B.4}$$
$$V_{BIAS1}=V_{DD}-V_{GT1}-\frac{A_{VF}}{2\cdot\frac{g_{m1}}{I_D}(V_{GT1},\theta)}$$

Equation (B.2) can be finally written as:

$$F_{BIAS}=1+\left\{\frac{NEF\cdot\dfrac{V_{GT1}}{V_{GT,BIAS1}(A_{VF},V_{DD},V_{GT1},\theta)}\cdot\dfrac{1+\frac{\theta}{2}\cdot V_{GT,BIAS1}(A_{VF},V_{DD},V_{GT1},\theta)}{1+\theta\cdot V_{GT,BIAS1}(A_{VF},V_{DD},V_{GT1},\theta)}\cdot\dfrac{1+\theta\cdot V_{GT1}}{1+\frac{\theta}{2}\cdot V_{GT1}}}{\dfrac{1}{(V_{DD}-V_{GT3}(A_{VF},V_{GT1},\theta)-V_{GT1})\cdot\frac{g_{m1}}{I_D}(V_{GT1},\theta)}}\right.$$

where $V_{GT,BIAS1}$ is the V_{GT} of the MOST used implement the current source I_{BIAS1}.

Appendix C
Two-port Amplifiers Noise and Power Matching

In this appendix, conditions for the noise-power match of a wide-band two-port amplifier are derived. The aim is to check when an amplifier can provide noise and power matching.

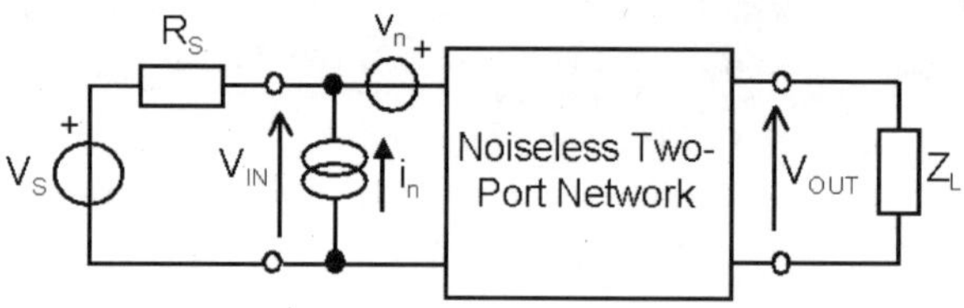

Figure 1: Two-port circuit noise model.

Consider the two-port noise model in figure 1, where the equivalent noise sources 'i_n' and 'v_n' are indeed the superposition of the contribution of M internal noise sources $I_{n,1}$, $I_{n,2}$, .. $I_{n,M}$ as: $i_n=i_{n,1}+i_{n,2}+..+i_{n,M}$ and $v_n=v_{n,1}+v_{n,2}+..+v_{n,M}$. According to noise-theory [1], the noise factor F of such a two-port can be written as:

$$F=1+\frac{\overline{\left(\sum_k^M i_{n,k}+G_S\cdot\sum_k^M v_{n,k}\right)^2}}{\overline{I_{n,S}^2}}=1+\frac{\overline{\left(\sum_k^M i_{n,U,k}+G_S\cdot\sum_k^M v_{n,k}+\sum_k^M G_{C,k}\cdot v_{n,k}\right)^2}}{\overline{I_{n,S}^2}} \qquad (C.1)$$

where the identity $i_{n,k}=i_{n,U,k}+i_{n,C,k}=i_{n,U,k}+G_{C,k}\cdot v_{n,k}$ has been used (with $G_{C,k}$ being the correlation conductance of the k-th noise source $v_{n,k}$). Equation (C.1) can be rewritten as:

$$F(G_S)=1+\frac{\overline{\left(\sum_k^M i_{n,U,k}+(G_S+G_C)\cdot\sum_k^M v_{n,k}\right)^2}}{\overline{I_{n,S}^2}}=1+\frac{\overline{i_{n,U}^2}+(G_S+G_C)^2\cdot\overline{v_n^2}}{\overline{I_{n,S}^2}} \qquad (C.2)$$

$$\overline{i_{n,U}^2}\equiv\sum_k^M\overline{i_{n,U,k}^2},\quad \overline{v_n^2}\equiv\sum_k^M\overline{v_{n,k}^2}\quad \text{and}\quad G_C\equiv\frac{\sum_k^M G_{C,k}\cdot v_{n,k}}{\sum_k^M v_{n,k}}$$

Equations (C.2) shows that the noise factor of the two-port in figure 1 can be expressed in terms of the quantities $i_{n,U}$, v_n and G_C if they are defined as above.
The two-port noise factor can be manipulated to provide [1]:

$$
\begin{aligned}
&F(G_S) = F_{MIN} + \frac{R_n}{G_S}\cdot(G_S - G_{S,OPT})^2 \\
&F_{MIN} = 1 + 2\cdot R_n\cdot(G_S + G_{S,OPT}) \\
&G_{S,OPT} = \sqrt{G_C^2 + \frac{G_U}{R_n}};\quad G_C = \frac{\overline{v_n\cdot i_n}}{\overline{v_n^2}} \quad\text{and}\quad G_U = \frac{\overline{(i_n - G_C\cdot v_n)^2}}{4kT\Delta f} \\
&R_n = \frac{\overline{v_n^2}}{4kT\Delta f}\quad G_S = \frac{\overline{I_{n,S}^2}}{4kT\Delta f}\quad G_U = \frac{\overline{i_{n,U}^2}}{4kT\Delta f}
\end{aligned}
\tag{C.4}
$$

Relations (C.4) show the existence of an optimum value of the source conductance, $G_{S,OPT}$, which provides the minimum value of the noise factor, F_{MIN} (with $G_S=1/R_S$ the conductance of the source). If $G_{S,OPT}$ is equal to G_{IN} and $G_{IN}=G_S$ holds, the two-port input is simultaneously optimised for noise and power transfer.
The (equivalent) noise sources $v_{n,k}$ and $i_{n,k}$ of the two-port in figure 1 are related to the output noise of the k-th device noise as (e.g.: using {A, B, C, D} parameters):

$$
\begin{aligned}
v_{n,k} &= V_{out,short,n,k}\cdot\frac{A\cdot Z_L + B}{Z_L} = I_{n,k}\cdot H_{n,short,k}\cdot\frac{A\cdot Z_L + B}{Z_L} \\
i_{n,k} &= V_{n,out,open,k}\cdot\frac{C\cdot Z_L + D}{Z_L} = I_{n,k}\cdot H_{n,open,k}\cdot\frac{C\cdot Z_L + D}{Z_L}
\end{aligned}
\tag{C.5}
$$

where $H_{n,short(open),k}$ is equal to the noise transfer function $H_{n,k}$ for the two-ports upon a shorted (open) input. From Equation (C.5), the equivalent sources $v_{n,k}$ and $i_{n,k}$ are *fully correlated*, because they are proportional to each other, thus $i_{n,k}=G_{C,k}\cdot v_{n,k}$ (i.e. $G_{U,k}$.=0) holds. Using equation (C.5), the correlation conductance $G_{C,k}$ can be written as:

$$
G_{C,k} \equiv \frac{i_{n,k}}{v_{n,k}} = \frac{H_{n,open,k}\cdot(C\cdot Z_L + D)}{H_{n,short,k}\cdot(A\cdot Z_L + B)} = \frac{H_{n,open,k}}{H_{n,short,k}}\cdot G_{IN}
\tag{C.6}
$$

From equation (C.4), the optimum conductance, $G_{S,OPT}$, of the source is indeed equal to:

$$G_{S,OPT} = G_C \equiv \frac{\sum_k^M G_{C,k} \cdot v_{n,k}}{\sum_k^M v_{n,k}} = \frac{\sum_k^M \frac{H_{n,open,k}}{H_{n,short,k}} \cdot G_{IN} \cdot I_{n,k} \cdot H_{n,short,k} \cdot \frac{A \cdot Z_L + B}{Z_L}}{\sum_k^M I_{n,k} \cdot H_{n,short,k} \cdot \frac{A \cdot Z_L + B}{Z_L}} \tag{C.7}$$

and finally:

$$G_{S,OPT} = G_C = G_{IN} \cdot \frac{\sum_k^M I_{n,k} \cdot H_{n,open,k}}{\sum_k^M I_{n,k} \cdot H_{n,short,k}} \tag{C.8}$$

Equation (C.8) shows as noise-power match (i.e. $G_{S,OPT}=G_{IN}$) yields to condition:

$$\frac{\sum_k^M I_{n,k} \cdot H_{n,open,k}}{\sum_k^M I_{n,k} \cdot H_{n,short,k}} = 1 \tag{C.9}$$

Equation (C.9) is true, if and only if relation $H_{n,open,k} = H_{n,short,k}$ holds $\forall$k. Using equation (C.4) and (C.8), the noise factor for $G_{IN}=G_S$ can be written as:

$$F = 1 + \frac{R_n}{R_S}\left[3 + \left(\frac{\sum_k^M I_{n,k} \cdot H_{n,open,k}}{\sum_k^M I_{n,k} \cdot H_{n,short,k}}\right)^2\right] \tag{C.10}$$

For $H_{n,open,k} = H_{n,short,k}$ equation (C.11) becomes:

$$F = F_{MIN} = 1 + \frac{4 \cdot R_n}{R_S} \tag{C.11}$$

From equation (C.11), the minimum noise factor is proportional to R_n/R_S.

References

[1] H. Rothe and W. Dahlke, "Theory of Noisy Four Poles", Proceedings IRE, vol.44, pp. 811-818, June 1956.

APPENDIX D: All 1- and 2-VCCS Graphs

In this appendix all 150 graphs with one or two VCCSs and their transmission parameters will be given. Also a few examples are given to show how graphs can be converted to circuits, so that readers can explore other circuits themselves.

GRAPHS WITH ONE VCCS

We will start with circuits with 1 VCCS. Such circuits consist of the elements shown in Figure 1: a signal source, one VCCS and a load impedance. Figure 1 also shows their graph representation.

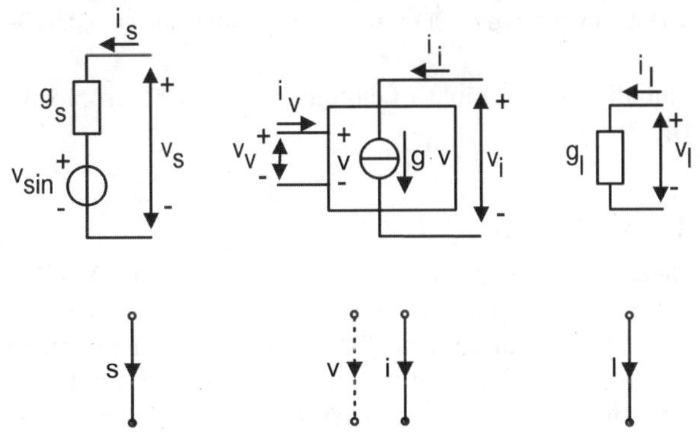

Figure 1: Circuit elements and their graph representations: an independent source (s-branch), a VCCS (v- and i-branch) and a load impedance (l-branch).

Figure 2 shows all graphs with one VCCS and at least one non-zero transmission parameters. Obviously it is possible to implement a transconductor with B=-1/g (e.g. with a common source transistor). Also, a series conductance and parallel conductance are possible (e.g. via a "self-connected" VCCS implementation). Finally for large transconductance values, a voltage follower can be approximated (only non-zero A=1, e.g. with a source follower) and a current follower (only non-zero D=1, e.g. a common gate stage). As expected, we appear to cover all the known single transistor amplifier stages.

The upper rightmost graph in Figure 2 is peculiar in the sense that it consists of two sub-graphs: a parallel s- and v- branch and a parallel i- and l-branch. These graphs can be joint at an arbitrary node indicated with dashed lines without changing the transmission parameters. This is because the VCCS v-branch senses the differential voltage across the source branch: the common mode voltage with respect to the load ideally has no effect. In practical circuits this is useful to implement a so-called "floating input", and the graphs will be labeled floating input graphs.

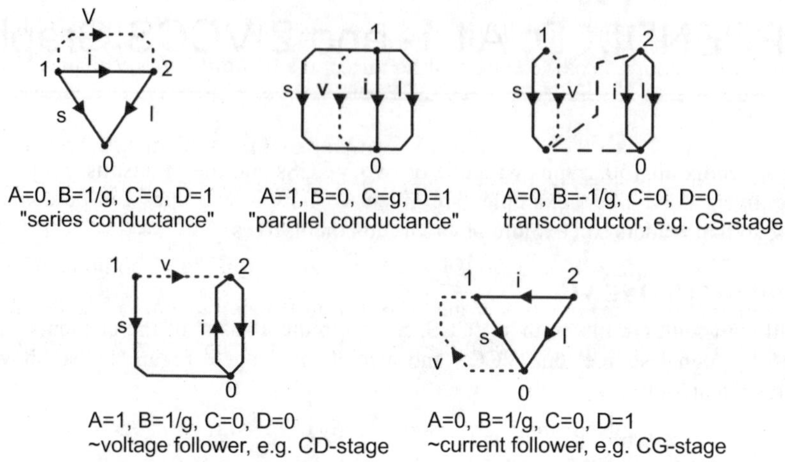

Figure 2: All potentially useful graphs of circuits with a source (s), load (l) and one VCCS (v and i) (see text).

GRAPHS WITH TWO VCCSs

From the experience gained in generating the graphs with one VCCS, rules for useless graphs have been inferred and proven. A very useful rather obvious rule states that v-branches do not play any role in Kirchhoff Current Law (KCL) relations (their current is always zero). Hence KCL relations can be derived from graphs without v-branches. As a consequence, it is possible to split the graph generation in two phases:

1. Generating all "Kirchhoff Current Law graphs" ("KCL-graphs") consisting of 2 i-branches (from 2 VCCSs) and a source (s) and load (l) branch (no v-branches). This is a task with limited complexity similar to that of the 1VCCS graph generation, which was done by hand.
2. Systematically adding v-branches to the graphs in all possible ways and analyzing the transfer function in terms of transmission parameters. This is a tedious job involving several hundreds of graphs performed by a MAPLE graph generation and symbolic analysis computer program.

We will indicate the two VCCSs with the index a and b. Figure 3 shows all KCL graphs that result from combining all possibilities and removing redundant equivalent ones (as the two VCCSs are equivalent, changing index from a to b results in an equivalent graph). The KCL graphs can consist of more pieces. They can also have a separate s-branch, just as in the transconductor graph with one VCCS in Figure 2.

For reasons of convenience, the KCL graphs have been given names, referring to the branch names and their interconnection structure: "+" refers to "in series with" and "//" to "in parallel with". Brackets are used to indicate groups.

To find all complete graphs, two v-branches are added systematically between all possible node-combinations. The separate s-branch is either left separate and tested for common mode insensitivity of the transfer function (for floating input graphs), or connected to each node of the rest of the graph (non-floating input graphs). This is done by introducing a special node "sref" (see the graphs in the low half of Figure 3). During the automated graph generation and analysis procedure, the sref-node is systematically connected to all other nodes in the graph. The "any" node of a separate i-branch acting as a norator could be handled in the same way. However, it can be shown that the analysis results are not affected by changing this node-connection, so one analysis for the graph is sufficient. Note that each of the graphs with an "any" node has several alternative implementations with the same transfer function. This will be exemplified in the voltage follower circuit generation given later in this appendix.

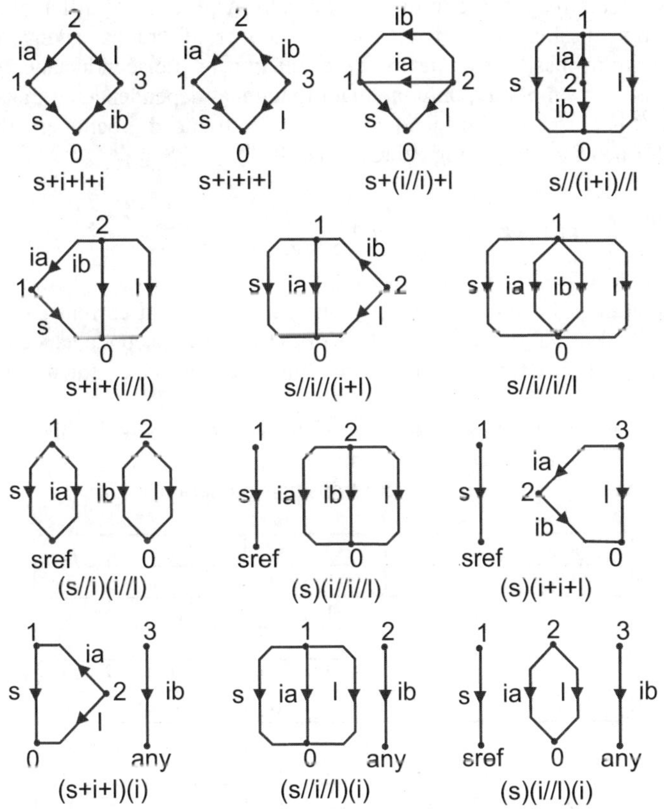

Figure 3: The 13 directed labeled Kirchhoff Current Law graphs for two-ports with two VCCSs. By adding va- and vb- branches systematically between all possible combinations of nodes, all graphs for two-ports with 2 VCCSs are generated.

TYPES OF POSSIBLE 2VCCS TWO-PORTS

The graph generation and analysis discussed in the previous paragraph leads to 145 graphs with at least one non-zero transmission parameter. They are listed at the end of this appendix, grouped on non-zero transmission parameters. As circuits with floating input are more flexible and have different applications, they are printed separately at the end (12 cases). For each two-port graph the following data is given:

- graph name referring to graphs in Figure 3
- Equations for the voltage va and vb and, if applicable, for the voltage on node sref. Node 0 is the reference node.
- Expressions for the transmission parameters [A,B,C,D] of the two-port.

Table 1 shows a summary of the results in Appendix A in terms of different combinations of non-zero transmission parameters, and the number of graphs having this property. Obviously, the transmission parameters are a function of transconductance ga and gb of the two VCCSs. For simplicity, only the main functional dependence on transconductance is shown. For this purpose a transconductance gi is introduced, i being an arbitrary index, that represents one of the following expressions:

$$g_i \in \left\{ \pm g_a, \pm g_b, \pm (g_a \pm g_b), \frac{\pm g_a g_b}{g_a \pm g_b} \right\} \qquad (1)$$

From considerations of dimension, the voltage gain (1/A) and current gain (1/D) of the two-port described in Table 1 are either 0, 1 or a ratio of transconductance expression gi. The transconductance (1/B) is equal to gi and transimpedance (1/C) equal to 1/gi.

Table 2 lists the 9 commonly desired types of linear two-ports with either very low, very high or accurate port impedances. The fourth column shows which two-port of Table 1 can be used to implement the desired non-zero transmission parameters.

Case	A	B	C	D	#
A	1	0	0	0	3
B	0	1/g1	0	0	37
D	0	0	0	1	3
AB	1 or g1/g2	1/g3	0	0	24
AD	1	0	0	1	6
BC	0	1/ g1	g2	0	2
BD	0	1/ g1	0	1 or g2/g3	24
ABC	1 or g1/g2	1/ g3	g4	0	3
ABD	1 or g1/g2	1/ g3	0	1 or g4/g5	24
ACD	1	0	g1	1	9
BCD	0	1/ g1	g2	1 or g3/g4	3
ABCD	1 or g1/g2	1/g3	g4	1 or g5/g6	7

Table 1: Different combinations of non-zero transmission parameters of 2-VCCS circuits. Expression gi is defined in equation (1), where i is an arbitrary index.

Although there are certainly limitations to what is possible (e.g.: there are only 2 degrees of freedom via ga and gb), it can be concluded that even with only two VCCSs many useful circuits can be implemented. As might be expected for circuits with VCCSs, cases with high and finite port impedances are readily available. For low port impedance, high transconductance is usually needed (except in cases where one VCCS acts as a nullor).
Care was taken to find equivalent circuit graphs by detecting the existence of equal sets of transmission parameters. In cases where this was observed, equivalent graphs have been rejected. Quite some graphs have transmission parameters that depend only on one transconductance. The other VCCS is "invisible" from the two-port parameters, suggesting that the other is not doing anything useful. However, this is not necessarily the case: it might for instance act as a nullor (e.g. in voltage followers below). In some other cases, the "invisible" VCCS is just an impedance or current follower in series with the current source of the "visible" VCCS. Assuming ideal VCCSs this is useless. However, in practical transistor circuits with only approximate VCCS behavior, it still can be useful. An example is a cascode transistor: it improves the output impedance but this is not evident from a circuit with ideal VCCSs! (they already have infinite output impedance). As our goal was to find all two-port circuits, we decided to maintain graphs in the database, unless we could prove that they are useless or identical to another one that is already there. In other words: we consider graphs "potentially useful", unless it is proven they are not.

Zin	Zout	Desired	Realizable with case	Additional conditions
∞	0	A	AB	B << A·Zl
∞	∞	B	B	-
0	0	C	BC	B << C·Zs·Zl
0	∞	D	BD	B << D·Zs
∞	= Zl	AB	AB	-
0	= Zl	CD	BCD	B << D·Zs
= Zs	0	AC	ABC	B << A·Zl
= Zs	∞	BD	BD	-
= Zs	= Zl	ABCD	ABCD	A·D = B·C

Table 2: Overview of the implementation possibilities of 9 desired two-ports using 2VCCS circuits.

FROM GRAPH TO CIRCUIT

Now we have a database of graphs of potentially useful circuits, the question is how to implement them on transistor level. Every VCCS in a graph can in principle be implemented using an arbitrary 4-terminal transconductor implementation. However, a single MOST or even a single resistor is sufficient in many cases. Whether this is possible depends on the interconnection and the orientation of the v- and i- branch of a VCCS in a graph. Figure 4 illustrates this point. From the figure we see that three different situation occur:

1. If there is no connection between the v- and i-branch, a VCCS with an isolated input and output ports is necessary (right most case). This can be implemented by common source MOST pairs, either of the same or of different type (biasing is not shown).
2. If one connection exists between the v- and i-branch, and the branches have the same orientation (both arrows pointing to or from the common node), a single MOST can be used (a PMOST or NMOST depending on the branch orientation). Alternatively, common source MOST pairs can be used.
3. If the branches are connected at both ends (left-most case), and also have the same orientation, then a simple resistor can be used, apart from single MOSTs or common source MOST pairs.

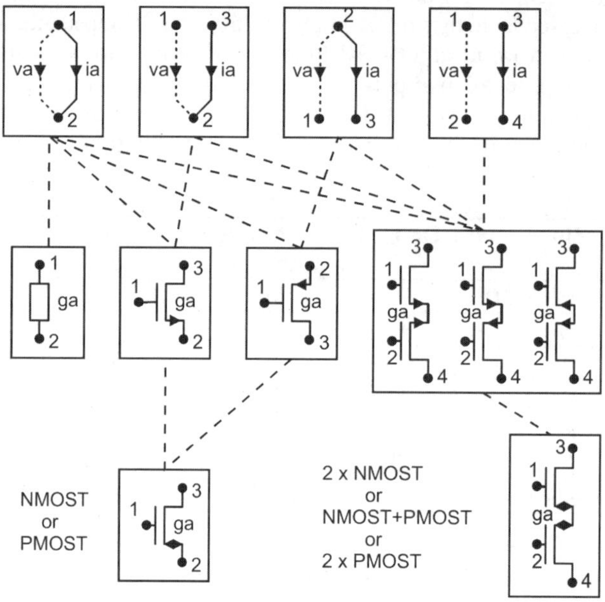

Figure 4: The interconnections between the v- and i-branches, and the relative orientation of the branches determines the possibilities to implement a VCCS. Dashed lines show alternative implementations. The double arrow MOST symbol refers to either NMOST or PMOST.

Thus it appears that branch direction has large impact on the implementation possibilities. Furthermore it has a strong impact on stability (e.g. reversal of the v-terminals may turn a self-connected positive impedance in a negative impedance as exploited in oscillators). Therefore, the direction of the branches will now be considered. The effect of v- or i-branch reversal is as follows:

- Change the direction of the v- or i-branch: sign change of the related transconductance-term in transmission parameters expressions.

- Change both the direction of the v- and i-branch of the same VCCS: no effect on the transfer function. However, according to Figure 4 this changes an NMOST into a PMOST and vice versa. This transformation has no effect on the transfer function, but can be useful for instance for biasing purposes to "fit" circuits in a low supply voltage.

As a PMOST implementation can always be replaced by an NMOST by changing both the v- and i- orientation without changing the transmission parameter equations, it is useful to have a symbol available which represents either a NMOST or a PMOST. The symbol that will be used in this paper for this purpose is a MOST symbol with double arrow, as shown at the bottom of Figure 4.

Note that it is important to indicate the source terminal, even though a MOS transistor physically is a symmetrical device. However, we assume it to implement a transconductor, which is NOT symmetrical anymore: the source is both a voltage-sense and current-source terminal, while the drain is only a current-source terminal.

As an example of the effect of branch reversal on the VCCS implementation possibilities consider Figure 5. It shows one of the (s)(i+i+l) transconductor graphs found in Appendix A. In Figure 5a the orientation of the v- and i-branches are different for VCCSa as well as for VCCSb. Therefore common source pairs have to be used for both ga and gb.

By changing the direction of the vb branch, vb and ib get the same orientation (Figure 5b). Since they are also in parallel, a simple resistor can be used. If the direction of ia is also changed (Figure 5c), VCCSa can be implemented by a single MOST, and the familiar source degenerated MOS circuit results.

$$B = \frac{g_a + g_b}{g_a g_b} \qquad B = \frac{-g_a + g_b}{g_a g_b} \qquad B = -\frac{g_a + g_b}{g_a g_b}$$

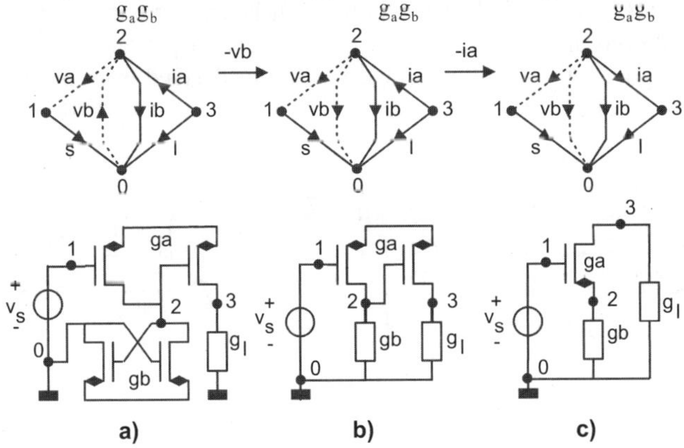

Figure 5: Example of the effect of va and vb sign changes on the transmission parameter expressions and the complexity of the circuit implementation (biasing not shown).

Because of these direction changes, transmission parameter B changes, as indicated above the graphs in Figure 5. Note that changing the direction of vb or ia results in a sign change of respectively gb and ga in B.

In this place, a note on stability is appropriate. Although the graph analysis renders solutions for the node voltages, it is not guaranteed that the solutions are stable. Many branch orientations render negative impedances, because of positive feedback loops. Therefore a careful stability consideration is needed, and positive overall node impedances should be guaranteed. By inspection of the graphs, the occurrence of a negative impedance can easily be recognized by qualitative reasoning. In Figure 5a, both VCCSa and VCCSb introduce a negative impedance (note the cross-coupled structure often used in latches). In Figure 5b only VCCSa does so, while case Figure 5c has all positive node impedances.

CIRCUIT EXAMPLES

To demonstrate the power of systematic circuit generation, and illustrate some previously mentioned issues, some examples of circuits found with the technique will be given. First current amplifier circuits will be discussed, and then voltage followers, as an example of circuits where one VCCS acts as a nullor.

Current Amplifiers

As discussed in section 0, in some cases the ideal desired combination of transmission parameters is not available. Still a practical useful approximation of the desired transfer function is often possible. Consider for example a current amplifier. Ideally this should only have non-zero D so that Zin is 0, Zout is infinite and the current gain is 1/D. Unfortunately only unity gain current amplifiers (current followers) are readily available. However, as shown in Table 2 an approximation with case BD is possible, provided that $B \ll D \cdot Zs$. Also case ABD with A=1 can be used, provided that it is acceptable that voltage changes at the output are copied to the input. When driving with a high ohmic current source these conditions are often satisfied. Alternatively large transconductance values for ga can be used. Figure 6 shows circuit implementations derived from the 3 BD and 3 ABD graphs with non-unity D-expressions.

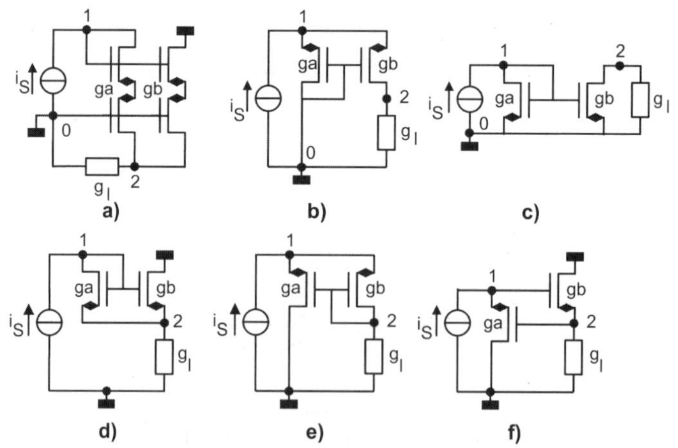

Figure 6: Current amplifier approximations derived from the BD graphs (a, b and c) and ABD (d, e, f) graphs with non-unity D. The current gain of the circuits is either higher than 1 (a, d), lower than 1 (b, e), or arbitrary (c,f).

Note that the ubiquitous current mirror is one of them (case c). This example demonstrates that even if the ideally desired two-port can not be implemented directly, practical useful approximations are often available.

Voltage Followers

In section 0 it was mentioned that one of the VCCSs may act as a nullor. This is the case for the unity gain voltage followers in Appendix A (A graphs, all of type (s)(i//l)(i)). These graphs can be jointly represented as shown in Figure 7. The nodes va_ref and ib_ref can be connected to node 0, 1 and 2 in 9 different combinations. VCCSa drives the load gl and VCCSb acts as a nullor (since current ib=0, voltage vb is also). This is illustrated in the right part of Figure 7.

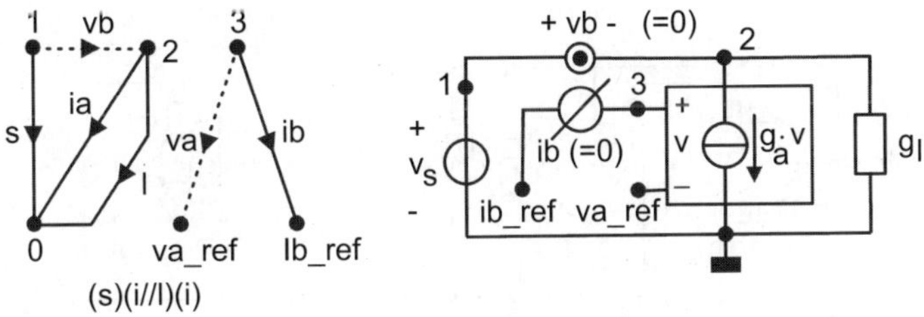

Figure 7: The voltage follower graph (s)(i//l)(i) in which VCCSb acts as a nullor, that forces the load voltage equal to the source voltage.

Although not necessary, suppose for simplicity that va_ref and ib_ref are grounded. Now, node 2 will follow voltage vs due to the nullator (vb branch), provided that ib acts as a norator providing negative feedback to the vb terminals. Indeed, VCCSa provides this feedback from node 3 (norator node) to node 2 (nullator node). If for instance vs increases, VCCSb will drive the voltage of node 3 down. As a result VCCSa drives additional current in the load impedance and the load voltage increases until vb is zero.
Working out all 9 different connections of node va_ref and ib_ref to other nodes in the circuit, 11 different voltage follower implementations can be derived as shown in Figure 8. This example clearly demonstrates the power of the systematic approach: many alternative circuits are generated that are not easily found otherwise.

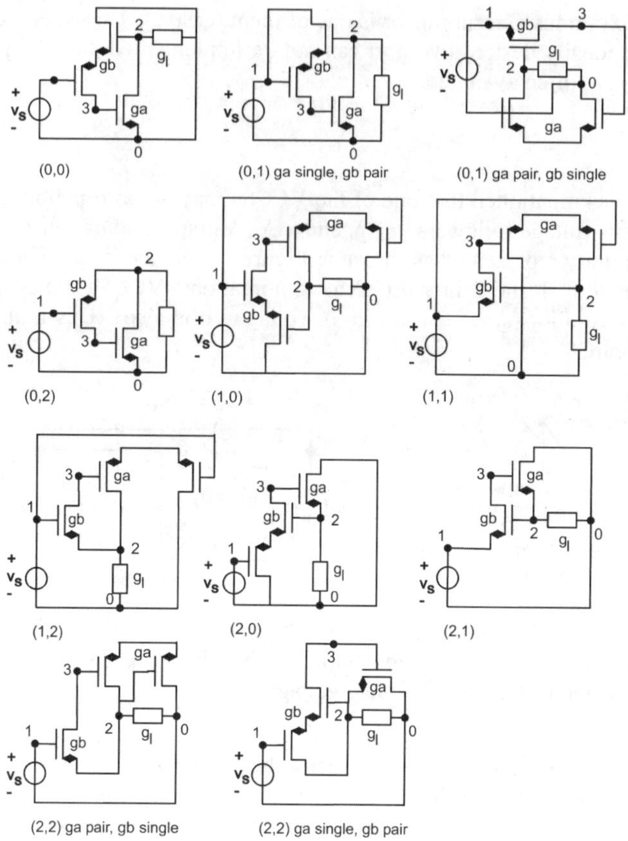

Figure 8: Voltage follower circuits derived from the graph in Figure 7 by systematically connecting the va_ref and ib_ref nodes to all other nodes in the circuits (their node numbers are indicated between brackets).

Lists of 1- and 2-VCCS Graphs

****
*****
NON-FLOATING INPUT GRAPHS
****
*****
THE 3 CASES WITH NON-ZERO PARAMETERS A:

(s)(i//l)(i) [va = v3, vb = v2-v1, sref = 0]
[A=1, B=0, C=0, D=0]
(s)(i//l)(i) [va = -v2+v3, vb = v2-v1, sref = 0]
[A=1, B=0, C=0, D=0]
(s)(i//l)(i) [va = v3-v1, vb = v2-v1, sref = 0]
[A=1, B=0, C=0, D=0]

****
*****
THE 28 CASES WITH NON-ZERO PARAMETERS B:

(s)(i//i//l) [va = v1, vb = v1, sref = 0]
[A=0, B=(-1/(ga+gb)), C=0, D=0]
(s)(i+i+l) [va = v2, vb = v2-v1, sref = 0]
[A=0, B=(-(-ga+gb)/gb/ga), C=0, D=0]
(s)(i+i+l) [va = v2-v1, vb = v2, sref = 0]
[A=0, B=((-ga+gb)/gb/ga), C=0, D=0]
(s)(i+i+l) [va = v1, vb = v2, sref = v2]
[A=0, B=(-(-ga+gb)/gb/ga), C=0, D=0]
(s)(i+i+l) [va = v2, vb = v1, sref = v2]
[A=0, B=((-ga+gb)/gb/ga), C=0, D=0]

(s)(i+i+l) [va = v1, vb = v2, sref = 0]
[A=0, B=(-1/ga), C=0, D=0]
(s)(i+i+l) [va = v1, vb = -v2+v3, sref = 0]
[A=0, B=(-1/ga), C=0, D=0]
(s)(i+i+l) [va = v1, vb = v2-v1, sref = 0]
[A=0, B=(-1/ga), C=0, D=0]
(s)(i+i+l) [va = v2, vb = v1, sref = 0]
[A=0, B=(-1/gb), C=0, D=0]
(s)(i+i+l) [va = -v2+v3, vb = v1, sref = 0]
[A=0, B=(-1/gb), C=0, D=0]
(s)(i+i+l) [va = v2-v1, vb = v1, sref = 0]
[A=0, B=(-1/gb), C=0, D=0]
(s)(i+i+l) [va = v1, vb = v2-v1, sref = v2]
[A=0, B=(1/gb), C=0, D=0]
(s)(i+i+l) [va = v2, vb = v2-v1, sref = v2]
[A=0, B=(1/gb), C=0, D=0]
(s)(i+i+l) [va = v3-v1, vb = v2-v1, sref = v2]
[A=0, B=(1/gb), C=0, D=0]
(s)(i+i+l) [va = -v2+v3, vb = v2-v1, sref = v2]
[A=0, B=(1/gb), C=0, D=0]
(s)(i//l)(i) [va = v3, vb = v3-v1, sref = 0]
[A=0, B=(-1/ga), C=0, D=0]
(s)(i//l)(i) [va = v3-v1, vb = v3, sref = 0]
[A=0, B=(1/ga), C=0, D=0]
(s)(i//l)(i) [va = v1, vb = v3, sref = 0]
[A=0, B=(-1/ga), C=0, D=0]
(s)(i//l)(i) [va = v1, vb = -v2+v3, sref = 0]
[A=0, B=(-1/ga), C=0, D=0]
(s)(i//l)(i) [va = v1, vb = v3-v1, sref = 0]
[A=0, B=(-1/ga), C=0, D=0]
(s)(i//l)(i) [va = v3, vb = v1, sref = v3]
[A=0, B=(1/ga), C=0, D=0]
(s)(i//l)(i) [va = -v2+v3, vb = v2-v1, sref = v3]
[A=0, B=(1/ga), C=0, D=0]
(s)(i//l)(i) [va = v2-v1, vb = -v2+v3, sref = v3]
[A=0, B=(1/ga), C=0, D=0]
(s)(i//l)(i) [va = v3-v1, vb = v3, sref = v3]
[A=0, B=(1/ga), C=0, D=0]
(s)(i//l)(i) [va = v3-v1, vb = -v2+v3, sref = v3]
[A=0, B=(1/ga), C=0, D=0]
(s)(i//l)(i) [va = v3-v1, vb = v2-v1, sref = v3]
[A=0, B=(1/ga), C=0, D=0]
(s)(i//l)(i) [va = v3-v1, vb = v1, sref = v3]
[A=0, B=(1/ga), C=0, D=0]
(s)(i//l)(i) [va = v1, vb = v3, sref = v3]
[A=0, B=(-1/ga), C=0, D=0]

****
*****
THE 3 CASES WITH NON-ZERO PARAMETERS D:

(s+i+l)(i) [va = v3-v1, vb = v1]
[A=0, B=0, C=0, D=1]
(s+i+l)(i) [va = -v2+v3, vb = v1]
[A=0, B=0, C=0, D=1]
(s+i+l)(i) [va = v3, vb = v1]
[A=0, B=0, C=0, D=1]

****
*****
THE 23 CASES WITH NON-ZERO PARAMETERS AB:
(s)(i//i//l) [va = v1, vb = v2, sref = 0]
[A=(-gb/ga), B=(-1/ga), C=0, D=0]
(s)(i//i//l) [va = v1, vb = v2-v1, sref = 0]
[A=(gb/(-ga+gb)), B=(1/(-ga+gb)), C=0, D=0]
(s)(i//i//l) [va = v2, vb = v2-v1, sref = 0]
[A=((ga+gb)/gb), B=(1/gb), C=0, D=0]
(s)(i//i//l) [va = v2-v1, vb = v2-v1, sref = 0]
[A=1, B=(1/(ga+gb)), C=0, D=0]
(s)(i+i+l) [va = -v2+v3, vb = v2-v1, sref = 0]
[A=1, B=(1/gb/ga*(ga+gb)), C=0, D=0]
(s)(i+i+l) [va = v2-v1, vb = -v2+v3, sref = 0]
[A=1, B=(1/gb/ga*(ga+gb)), C=0, D=0]
(s)(i+i+l) [va = v1, vb = -v2+v3, sref = v2]
[A=-1, B=(-1/gb/ga*(ga+gb)), C=0, D=0]
(s)(i+i+l) [va = v3-v1, vb = v2, sref = v2]
[A=1, B=(1/gb/ga*(ga+gb)), C=0, D=0]

(s)(i+i+l) [va = v2, vb = v3-v1, sref = 0]
[A=1, B=(1/gb), C=0, D=0]
(s)(i+i+l) [va = v3-v1, vb = v2, sref = 0]
[A=1, B=(1/ga), C=0, D=0]
(s)(i+i+l) [va = v3-v1, vb = -v2+v3, sref = 0]
[A=1, B=(1/ga), C=0, D=0]
(s)(i+i+l) [va = v3-v1, vb = v2-v1, sref = 0]
[A=1, B=(1/ga), C=0, D=0]
(s)(i+i+l) [va = -v2+v3, vb = v3-v1, sref = 0]
[A=1, B=(1/gb), C=0, D=0]
(s)(i+i+l) [va = v2-v1, vb = v3-v1, sref = 0]
[A=1, B=(1/gb), C=0, D=0]
(s)(i//l)(i) [va = -v2+v3, vb = v3-v1, sref = 0]
[A=1, B=(-1/ga), C=0, D=0]
(s)(i//l)(i) [va = v2-v1, vb = v3, sref = 0]
[A=1, B=(1/ga), C=0, D=0]
(s)(i//l)(i) [va = v2-v1, vb = -v2+v3, sref = 0]
[A=1, B=(1/ga), C=0, D=0]
(s)(i//l)(i) [va = v2-v1, vb = v3-v1, sref = 0]
[A=1, B=(1/ga), C=0, D=0]
(s)(i//l)(i) [va = v3-v1, vb = -v2+v3, sref = 0]
[A=1, B=(1/ga), C=0, D=0]
(s)(i//l)(i) [va = v3, vb = v2-v1, sref = v3]
[A=1, B=(1/ga), C=0, D=0]
(s)(i//l)(i) [va = -v2+v3, vb = v1, sref = v3]
[A=-1, B=(1/ga), C=0, D=0]
(s)(i//l)(i) [va = v2-v1, vb = v3, sref = v3]
[A=1, B=(1/ga), C=0, D=0]
(s)(i//l)(i) [va = v1, vb = -v2+v3, sref = v3]
[A=-1, B=(-1/ga), C=0, D=0]

****
*****

****
THE 6 CASES WITH NON-ZERO PARAMETERS AD:

s//(i+i)//l [va = v2, vb = v2]
[A=1, B=0, C=0, D=1]
(s//i//l)(i) [va = v2, vb = v2]
[A=1, B=0, C=0, D=1]
(s//i//l)(i) [va = v2-v1, vb = v2-v1]
[A=1, B=0, C=0, D=1]
(s+i+l)(i) [va = v3-v1, vb = v2-v1]
[A=1, B=0, C=0, D=1]
(s+i+l)(i) [va = -v2+v3, vb = v2-v1]
[A=1, B=0, C=0, D=1]
(s+i+l)(i) [va = v3, vb = v2-v1]
[A=1, B=0, C=0, D=1]

****
THE 1 CASES WITH NON-ZERO PARAMETERS BC:

(s//i)(i//l) [va = v2, vb = v1, sref = 0]
[A=0, B=(-1/gb), C=ga, D=0]

****
THE 23 CASES WITH NON-ZERO PARAMETERS BD:

s+i+l+i [va = v2, vb = v2-v1]
[A=0, B=(-1/gb/ga*(ga+gb)), C=0, D=1]
s+i+l+i [va = v2-v1, vb = v2]
[A=0, B=(1/gb/ga*(ga+gb)), C=0, D=1]
s+i+i+l [va = v2-v1, vb = v2]
[A=0, B=((-ga+gb)/gb/ga), C=0, D=1]
s+i+i+l [va = v2, vb = v2-v1]
[A=0, B=(-(-ga+gb)/gb/ga), C=0, D=1]
s+(i//i)+l [va = v1, vb = v1]
[A=0, B=(1/(ga+gb)), C=0, D=1]
s+i+(i//l) [va = v1, vb = v1]
[A=0, B=(-1/(ga+gb)), C=0, D=(1/(ga+gb)*ga)]
s//i//(i+l) [va = v1, vb = v1]
[A=0, B=(-1/gb), C=0, D=((-ga+gb)/gb)]
(s//i)(i//l) [va = v1, vb = v1, sref = 0]
[A=0, B=(-1/gb), C=0, D=(-ga/gb)]

s+i+l+i [va = v1, vb = v2]
[A=0, B=(-1/ga), C=0, D=1]
s+i+l+i [va = v1, vb = v3]
[A=0, B=(-1/ga), C=0, D=1]
s+i+l+i [va = v1, vb = v2-v1]
[A=0, B=(-1/ga), C=0, D=1]
s+i+l+i [va = v1, vb = v3-v1]
[A=0, B=(-1/ga), C=0, D=1]
s+i+i+l [va = -v2+v3, vb = v1]
[A=0, B=(-1/gb), C=0, D=1]
s+i+i+l [va = v2-v1, vb = v1]
[A=0, B=(-1/gb), C=0, D=1]
s+i+i+l [va = v2, vb = v1]
[A=0, B=(-1/gb), C=0, D=1]
s+i+i+l [va = v1, vb = -v2+v3]
[A=0, B=(-1/ga), C=0, D=1]
s+i+i+l [va = v1, vb = v2-v1]
[A=0, B=(-1/ga), C=0, D=1]
s+i+i+l [va = v1, vb = v2]
[A=0, B=(-1/ga), C=0, D=1]
(s+i+l)(i) [va = v3-v1, vb = v3]
[A=0, B=(1/ga), C=0, D=1]
(s+i+l)(i) [va = v3, vb = v3-v1]
[A=0, B=(-1/ga), C=0, D=1]
(s+i+l)(i) [va = v1, vb = v3-v1]
[A=0, B=(-1/ga), C=0, D=1]
(s+i+l)(i) [va = v1, vb = -v2+v3]
[A=0, B=(-1/ga), C=0, D=1]
(s+i+l)(i) [va = v1, vb = v3]
[A=0, B=(-1/ga), C=0, D=1]

****
THE 3 CASES WITH NON-ZERO PARAMETERS ABC:

s+i+(i//l) [va = v2, vb = v2-v1]
[A=((ga+gb)/gb), B=(1/gb), C=(-ga), D=0]
s+i+(i//l) [va = v2, vb = v1]
[A=(-ga/gb), B=(-1/gb), C=(-ga), D=0]
(s//i)(i//l) [va = v2, vb = v2-v1, sref = 0]
[A=1, B=(1/gb), C=ga, D=0]

****
THE 24 CASES WITH NON-ZERO PARAMETERS ABD:
s+i+l+i [va = v2, vb = v3-v1]
[A=-1, B=(-1/gb/ga*(ga+gb)), C=0, D=1]
s+i+l+i [va = v3, vb = v2-v1]
[A=1, B=(-1/gb/ga*(ga+gb)), C=0, D=1]
s+i+l+i [va = v2-v1, vb = v3]
[A=1, B=(1/gb/ga*(ga+gb)), C=0, D=1]
s+i+l+i [va = v3-v1, vb = v2]
[A=-1, B=(1/gb/ga*(ga+gb)), C=0, D=1]
s+i+i+l [va = -v2+v3, vb = v2-v1]
[A=1, B=(1/gb/ga*(ga+gb)), C=0, D=1]
s+i+i+l [va = v2-v1, vb = -v2+v3]
[A=1, B=(1/gb/ga*(ga+gb)), C=0, D=1]
s+(i//i)+l [va = v2, vb = v2-v1]
[A=((ga+gb)/gb), B=(-1/gb), C=0, D=1]
s+(i//i)+l [va = v2, vb = v1]
[A=(-ga/gb), B=(1/gb), C=0, D=1]
s+(i//i)+l [va = v2-v1, vb = v2-v1]
[A=1, B=(-1/(ga+gb)), C=0, D=1]
s+(i//i)+l [va = v2-v1, vb = v1]
[A=(-1/(-ga+gb)*ga), B=(1/(-ga+gb)), C=0, D=1]
s+i+(i//l) [va = v2-v1, vb = v2-v1]
[A=1, B=(1/(ga+gb)), C=0, D=(1/(ga+gb)*ga)]
s//i//(i+l) [va = v2-v1, vb = v2-v1]
[A=1, B=(1/gb), C=0, D=((-ga+gb)/gb)]
(s//i)(i//l) [va = v2-v1, vb = v2-v1, sref = 0]
[A=1, B=(1/gb), C=0, D=(-ga/gb)]

s+i+i+l [va = -v2+v3, vb = v3-v1]
[A=1, B=(1/gb), C=0, D=1]
s+i+i+l [va = v3-v1, vb = -v2+v3]
[A=1, B=(1/ga), C=0, D=1]
s+i+i+l [va = v3-v1, vb = v2-v1]
[A=1, B=(1/ga), C=0, D=1]
s+i+i+l [va = v3-v1, vb = v2]
[A=1, B=(1/ga), C=0, D=1]
s+i+i+l [va = v2-v1, vb = v3-v1]
[A=1, B=(1/gb), C=0, D=1]
s+i+i+l [va = v2, vb = v3-v1]
[A=1, B=(1/gb), C=0, D=1]
(s+i+l)(i) [va = v3-v1, vb = -v2+v3]
[A=1, B=(1/ga), C=0, D=1]
(s+i+l)(i) [va = -v2+v3, vb = v3-v1]
[A=1, B=(-1/ga), C=0, D=1]
(s+i+l)(i) [va = v2-v1, vb = v3-v1]
[A=1, B=(1/ga), C=0, D=1]
(s+i+l)(i) [va = v2-v1, vb = -v2+v3]
[A=1, B=(1/ga), C=0, D=1]
(s+i+l)(i) [va = v2-v1, vb = v3]
[A=1, B=(1/ga), C=0, D=1]

****
THE 9 CASES WITH NON-ZERO PARAMETERS ACD:

s//(i+i)//l [va = v2, vb = v2-v1]
[A=1, B=0, C=(-gb*ga/(ga+gb)), D=1]
s//(i+i)//l [va = v2-v1, vb = v2]
[A=1, B=0, C=(gb*ga/(ga+gb)), D=1]

s//i//i//l [va = v1, vb = v1]
 [A=1, B=0, C=(ga+gb), D=1]

s//(i+i)//l [va = v1, vb = v2]
 [A=1, B=0, C=(-ga), D=1]
s//(i+i)//l [va = v1, vb = v2-v1]
 [A=1, B=0, C=(-ga), D=1]
(s//i//l)(i) [va = v1, vb = v2]
 [A=1, B=0, C=ga, D=1]
(s//i//l)(i) [va = v1, vb = v2-v1]
 [A=1, B=0, C=ga, D=1]
(s//i//l)(i) [va = v2, vb = v2-v1]
 [A=1, B=0, C=ga, D=1]
(s//i//l)(i) [va = v2-v1, vb = v2]
 [A=1, B=0, C=(-ga), D=1]

****
*****
THE 3 CASES WITH NON-ZERO PARAMETERS BCD:
s//i//(i+l) [va = v2, vb = v1]
 [A=0, B=(-1/gb), C=ga, D=1]
s//i//(i+l) [va = v2-v1, vb = v1]
 [A=0, B=(-1/gb), C=ga, D=((ga+gb)/gb)]
(s//i)(i//l) [va = v2-v1, vb = v1, sref = 0]
 [A=0, B=(-1/gb), C=ga, D=(ga/gb)]

****
*****
THE 7 CASES WITH NON-ZERO PARAMETERS ABCD:
s+i+(i//l) [va = v2-v1, vb = v1]
 [A=-ga/(-ga+gb), B=-1/(-ga+gb), C=-ga*gb/(-ga+gb), D=-ga/(-ga+gb)]
s+i+(i//l) [va = v2-v1, vb = v2]
 [A=((ga+gb)/ga), B=(1/ga), C=gb, D=1]
s+i+(i//l) [va = v1, vb = v2-v1]
 [A=gb/(-ga+gb), B=1/(-ga+gb), C=-ga*gb/(-ga+gb), D=- ga/(-ga+gb)]
s+i+(i//l) [va = v1, vb = v2]
 [A=(-gb/ga), B=(-1/ga), C=gb, D=1]
s//i//(i+l) [va = v1, vb = v2-v1]
 [A=1, B=(1/gb), C=ga, D=((ga+gb)/gb)]
s//i//(i+l) [va = v2, vb = v2-v1]
 [A=1, B=(1/gb), C=ga, D=1]
(s//i)(i//l) [va = v1, vb = v2-v1, sref = 0]
 [A=1, B=(1/gb), C=ga, D=(ga/gb)]

****
*****
END OF NON-FLOATING INPUT CASES (133 CASES)
****
*****

****
*****
FLOATING INPUT GRAPHS
****
*****
THE 9 CASES WITH NON-ZERO PARAMETERS B:

(s)(i//i//l) [va = v3-v1, vb = v3-v1]
 [A=0, B=(1/(ga+gb)), C=0, D=0]
(s)(i+i+l) [va = v4-v2, vb = v2-v1]
 [A=0, B=(1/gb/ga*(ga+gb)), C=0, D=0]

(s)(i+i+l) [va = v3-v2, vb = v4-v1]
 [A=0, B=(1/gb), C=0, D=0]
(s)(i+i+l) [va = v4-v2, vb = v4-v1]
 [A=0, B=(1/gb), C=0, D=0]
(s)(i+i+l) [va = v2, vb = v4-v1]
 [A=0, B=(1/gb), C=0, D=0]
(s)(i//l)(i) [va = v3-v1, vb = v4-v3]
 [A=0, B=(1/ga), C=0, D=0]
(s)(i//l)(i) [va = v4-v1, vb = v3]
 [A=0, B=(1/ga), C=0, D=0]
(s)(i//l)(i) [va = v4-v1, vb = v3-v1]
 [A=0, B=(1/ga), C=0, D=0]
(s)(i//l)(i) [va = v4-v1, vb = v3-v2]
 [A=0, B=(1/ga), C=0, D=0]

****
*****
THE 1 CASES WITH NON-ZERO PARAMETERS AB:

(s)(i//i//l) [va = v2, vb = v3-v1]
 [A=(ga/gb), B=(1/gb), C=0, D=0]

****
*****
THE 1 CASES WITH NON-ZERO PARAMETERS BC:

(s//i)(i//l) [va = v2, vb = v3-v1]
 [A=0, B=(1/gb), C=ga, D=0]

****
*****
THE 1 CASES WITH NON-ZERO PARAMETERS BD:
(s//i)(i//l) [va = v3-v1, vb = v3-v1]
 [A=0, B=(1/gb), C=0, D=(-ga/gb)]

****
****
END OF FLOATING INPUT CASES (12 CASES)
****
****

Index

S

T

U

V

W

Author Biographies

Federico Bruccoleri was born on February 21st, 1970, in Agrigento, Italy. In 1995 he received the M.Sc. degree in electrical engineering from the University of Genoa, Italy. In December 1997 he joined the IC-Design group of the MESA+ Research Institute, University of Twente, Enschede, The Netherlands where he receives his Ph.D. degree in November 2003. He is currently with Catena Microelectronics, Delft, The Netherlands. His research interests include the systematic design of high frequency linear CMOS circuits for Telecom applications. He received the "Best Poster Paper" award of the 2000 European Solid State Circuits Conference (ESSCIRC) and the Jan Van Vessen Award for Outstanding European paper of the IEEE 2002 International Solid-State Circuit Conference (ISSCC).

Eric A. M. Klumperink was born on April 4th, 1960, in Lichtenvoorde, The Netherlands. He received the B.Sc. degree from HTS, Enschede, The Netherlands, in 1982. After a short period in industry, he joined the Faculty of Electrical Engineering of the University of Twente in 1984, where he was mainly engaged in analog CMOS circuit design. This resulted in several publications and a Ph.D. thesis, in 1997, on the subject of "Transconductance Based CMOS Circuits". He is currently an Assistant Professor at the IC-Design Laboratory and also involved in the MESA+ Research Institute. He holds 4 patents and authored and co-authored more than 50 journal and conference papers. His research interest is in design issues of HF CMOS circuits, especially for front ends of integrated CMOS transceivers. He is co-recipient of the ISSCC 2002 "Van Vessem Outstanding Paper Award".

Bram Nauta was born in Hengelo, The Netherlands, in 1964. In 1987 he received the M.Sc degree (cum laude) in electrical engineering from the University of Twente, Enschede, The Netherlands. In 1991 he received the Ph.D. degree from the same university on the subject of analog CMOS filters for very high frequencies.

In 1991 he joined the Mixed-Signal Circuits and Systems Department of Philips Research, Eindhoven the Netherlands, where he worked on high speed AD converters. From 1994 he led a research group in the same department, working on "analog key modules". In 1998 he returned to the University of Twente, as full professor heading the department of IC Design group in the MESA+ Research Institute and department of Electrical Engineering. His current research interest is analog CMOS circuits for transceivers. Besides, he is also part-time consultant in industry and in 2001 he co-founded Chip Design Works.

His Ph.D. thesis was published as a book: *Analog CMOS Filters for Very High Frequencies*, Boston, MA, Kluwer, 1993. He holds 10 patents in circuit design and he received the "Shell Study Tour Award" for his Ph.D. Work. From 1997-1999 he served as Associate Editor of IEEE Transactions on Circuits and Systems -II; Analog and Digital Signal Processing, and in 1998 he served as Guest Editor for IEEE Journal of Solid-State Circuits. In 2001 he became Associate Editor for IEEE Journal of Solid-State Circuits. He is member of the technical program committee of ESSCIRC and ISSCC. He is co-recipient of the ISSCC 2002 "Van Vessem Outstanding Paper Award".